THE MOLECULAR BASIS
OF
HEREDITY

THE MOLECULAR BASIS
OF
HEREDITY

A. R. PEACOCKE, M.A., D.Sc.

Fellow and Tutor in Physical Biochemistry,
St. Peter's College, Oxford

and

R. B. DRYSDALE, B.Sc., Ph.D.

Lecturer in Microbiology, the University of Birmingham

REVISED REPRINT

SPRINGER SCIENCE+BUSINESS MEDIA, LLC

First published by
Butterworth & Co. (Publishers) Ltd.

First Edition, 1965
Revised and Reprinted, 1967

© Springer Science+Business Media New York 1965
Ursprünglich erschienen bei Butterworth & Co. (Publishers), Ltd. 1965

ISBN 978-1-4899-6163-1   ISBN 978-1-4899-6317-8 (eBook)
DOI 10.1007/978-1-4899-6317-8

Suggested **U.D.C.** *number*—575.1 : 547.963.32

CONTENTS

v

PART 3: THE STRUCTURE OF THE NUCLEIC ACIDS IN RELATION TO THEIR BIOLOGICAL FUNCTION

PART 4: SYNOPSIS AND ADDENDUM

PREFACE

The two intertwined chains of the molecule of deoxyribonucleic acid have come to represent to many, both scientists and non-scientists, a new phase in man's understanding of the nature of life. It has, for example, appropriately been used as a visual symbol for modern biochemistry at the most recent international congress devoted to this subject. But its suitability as a symbol may well be thought to go beyond its immediate molecular reference, for our knowledge of the genetic substance is itself the result of the intricate intertwining of a wide range of disciplines which may broadly be described as chemistry and biology, the structural and the functional.

The work which follows is the result of co-operation between two authors of whom one has been concerned with the structure and physical chemistry of nucleic acids and the other with genetics and microbiology. The mutual dialogue which was necessary for exposition of the molecular basis of heredity frequently proved how difficult it is for the terms of one discipline to be understood readily and employed accurately by someone trained in another. The main purpose of the ensuing text is to provide an account of the molecular basis of heredity suitable for undergraduates in their Final Honours work and for those beginning research in this field, but it is hoped that it will also contribute to that dialogue between different scientific disciplines which is so essential for the deeper understanding of living matter. The wider medical, social, and indeed, philosophical implications of this new insight into the chemical basis of heredity are beyond the scope of this book, but the knowledge which it describes should provide the factual basis for any other reflections of wider import.

The primary emphasis is on the structural aspect but it is the intention that the biological relevance should be depicted sufficiently fully for the reader, whether physical or biological scientist, to assess the significance of the structures described. The first part of the book (Chapters 1-3) gives an account of the earlier background to our present knowledge of the molecular basis of heredity and of the observations in which the nucleic acids were first identified as the chemical carriers of genetic information. In the second part (Chapters 4-7) a more detailed exposition is given of the chemical structure of these macromolecules and of the way in which they are constituted as the chromosomes. The third section (Chapters 9–12) describes the biological function of the nucleic acids in relation to their structure, a field of enquiry in which current activity is very intense. The addendum indicates briefly some of the most recent developments which have occurred since the main chapters were written.

The authors intended to confine references at the end of each chapter to a short list of pertinent reviews: in the event this proved impossible since so many of the important developments had at this stage been reported only in the original papers. The references are meant to serve only as a guide to the literature and are not comprehensive.

A glossary of a number of biological terms, some well established and some recent, is provided as a guide for those less familiar with biological usage; in the nature of the case it can be neither complete nor fully explicit.

This work developed out of an earlier survey in *Biological Reviews* in 1961, which was itself an attempt to bring up to date an article in *Endeavour* in 1957 by one of the present authors in conjunction with Professor W. G. Overend, of Birkbeck College, University of London. The increasing length of these successive efforts is symptomatic both of the extremely rapid growth in knowledge and of the problems which this growth poses for author, reader and investigator.

We are grateful to the many authors and publishers who gave permission for their diagrams to be reproduced. Individual acknowledgement is briefly, though inadequately, made at the appropriate point. We were glad to have the help of Mr. R. Mayne, of St. Peter's College, Oxford, in checking many details in the text and references. We are particularly conscious of our debt to the staff of the publishers for their patience and skill in bringing the work to print.

A. R. PEACOCKE
R. B. DRYSDALE

December 1964

PART 1
THE BACKGROUND

CHAPTER 1

INTRODUCTION

GENETICS[1]

Practical knowledge of heredity is older than the written word but the development of modern genetics dates only from the rediscovery of Mendel's paper in 1900, 34 years after it was published. During this period Mendel's work made little impression on biologists but the developments which did occur in biology made for a more ready acceptance of Mendelian theory when de Vries, Correns and Tschermark, probably independently, rescued Mendel's paper from oblivion.

Before 1900 the concept of the nucleus as the part of the cell responsible for heredity was suggested by Haeckel and developed by Hertwig and Strasberger. At the same time accurate descriptions of fertilization and of the multiplication of chromosomes by longitudinal splitting were produced. About 1885, Weismann pointed out that the germ cells of each generation are descended directly from the germ cells of the previous generation and not from the specialized body cells of the organism.

Weismann also emphasized the importance of the constancy of chromosome numbers and that the number must be halved during the formation of eggs and sperm. This prediction was confirmed by the observations of Boveri who thus established the double nature of the somatic cells as compared with the single nature of the gametes.

In his paper published in 1866, Mendel described his experiments in crossing edible peas and the conclusions he had reached. The basic difference between Mendel's work and that of his many predecessors who tried to discover the laws of heredity was that Mendel studied the inheritance of only a few characters, examined all the individuals of each generation and counted the numbers which showed the different characteristics, whereas his predecessors used varieties differing in large numbers of characters and looked for overall qualitative differences. He showed that the characters present in a hybrid segregated according to a simple mathematical ratio in the following generation, which was derived by self-fertilizing each hybrid plant. At first, he studied single pairs of contrasting characters (for example, full versus shrunken seeds, yellow versus green seeds, tall versus dwarf

3

plants and white versus coloured flowers) and, on the basis of his results, suggested that the factors determining the characters segregate during gamete production in the hybrid and are recovered unchanged in the following generation. This has become known as Mendel's first law. He also examined the behaviour of hybrids in which two or even three pairs of contrasting characters were segregating and found that each of the pairs of characters segregated independently. The principle of independent assortment is known as Mendel's second law.

In 1900 Mendel's results and their general application were confirmed by de Vries, Correns and Tschermak thereby marking the beginning of modern genetics. Soon afterwards, in 1903, Sutton gave a clear account of the relation between genes and chromosomes and provided a sound cytological basis for Mendel's two principles of segregation and independent assortment. From this, and similar independent work by Correns and Boveri, the chromosome theory of inheritance developed and finally was proved convincingly by Bridges' study[2] of the genetics and cytology of non-disjunction, in which both X chromosomes in the female Drosophila go to the same pole at meiosis.

The pairs of characteristics which Mendel studied all recombined independently, but after 1900 it became apparent that in some organisms there were more pairs of genes than pairs of chromosomes. This apparent contradiction was resolved by Morgan's theory of linkage which suggested that a number of genes may be located on one chromosome and that the strength of the linkage between them is inversely proportional to the probability that crossing over will occur between them. Sturtevant carried this a step further by suggesting a linear arrangement of genes in the chromosome and produced the first chromosome map.

After Bridges' paper in 1916, the chromosome theory of inheritance was not seriously challenged and the emphasis in genetic research shifted to problems such as the inheritance of continuously variable characters, the development of statistical methods, evolution, the genetics of development and mutation.

The rapid development of modern biochemical genetics, that is the study of gene structure and gene function at a biochemical level, dates from the work of Beadle and Tatum[3] on nutritionally exacting strains of the fungus *Neurospora crassa* and the demonstration by Avery, MacLeod and McCarty[4] that the genetic material is deoxyribonucleic acid (DNA). During the last 20 years, mainly through developments in microbial genetics, it has become firmly established that the function of genes is to control the structure or synthesis of proteins (*see* Chapters 10 and 11).

NUCLEIC ACIDS[5, 6]

Friederich Miescher in 1868 undertook an examination of pus cells in order to determine the chemical nature of their nuclear material. The cells were treated with pepsin and hydrochloric acid, which digested the cytoplasm leaving the nuclei as an insoluble mass. When this mass was extracted with sodium carbonate and the resulting solution acidified, a flocculent precipitate was obtained. Miescher regarded this substance as the characteristic constituent of the nucleus and suggested the term 'nuclein' for it. The properties of this substance were unusual. It was more acidic than proteins, insoluble in dilute acid, soluble in dilute alkali and contained considerable amounts of phosphorus but no sulphur. Miescher sent a manuscript with his remarkable results to Hoppe-Seyler who repeated Miescher's work with pus cells and also isolated a nuclein from yeast and in 1871 published the original manuscript and the confirmation of it in *Hoppe-Seyler's Medizinische-Chemische Untersuchunger*.

At about this time Miescher's interest shifted from pus cells to salmon sperm, a material which was readily available from salmon caught in the Rhine. Miescher found that the sperm heads were made up almost exclusively of a single chemical substance, a nuclein, which was a salt of an organic acid containing phosphorus and a basic protein which he called 'protamine'. The term 'nucleic acid' was introduced by Altmann in 1889 to describe the protein-free organic acids obtainable from nuclei. In the same paper, Altmann described the first general method for preparing protein-free nucleic acids from animal tissues and yeast. During the next 40 years the newer methods of nucleic extraction which were developed were based on the method of Altmann. However, following the revival of interest in nucleic acids in the early forties new, improved, methods for preparing nucleic acids were developed.

Following the publication of Altmann's method for the preparation of protein-free nucleic acids, Kossel studied the hydrolysis products of such preparations and laid the foundations of the organic chemistry of nucleic acids. While the earlier work on the components of the nucleic acids was qualitative in nature, by 1910 it was accepted that 'thymonucleic acid' contained two purines, adenine and guanine, and two pyrimidines, thymine and cytosine, all in equimolar amounts. By the same date it was also regarded as established that plant nucleic acids contained equimolar quantities of adenine, guanine, cytosine and uracil though only two plants, yeast and wheat, had been studied.

The nature of the sugar component of the nucleic acids was not finally settled until 1929 when it became clear that D-deoxyribose was the sugar component in thymonucleic acid. D-Ribose had been identified in plant nucleic acid preparations nearly twenty years earlier than this.

For the next decade it was taken for granted that there were two types of nucleic acid, one characteristic of animals and the other found only in plants. Additionally it was assumed that the nucleic acids had a tetranucleotide structure. The tetranucleotide theory became the 'central dogma' of its day, even though contradictory evidence was already available.

In the early 1940's, the chemistry and biology of nucleic acids was set on a new foundation by the development of new isolation and analytical techniques which were used to demonstrate that both types of nucleic acid were to be found within both plant and animal cells and that the molar proportions of the bases in nucleic acids did not agree with those required by the tetranucleotide theory[7, 8]. This new information coincided with the demonstration of the biological importance of nucleic acids and resulted in a vast and continuing expansion in nucleic acid research which gained fresh impetus with the proposal (Chapter 4) that deoxyribonucleic acid had a double-helical structure. This book describes some of the advances which have resulted from this coming together of chemistry and biology in elucidating the molecular basis of heredity.

REFERENCES

[1] Sturtevant, A. H. and Beadle, G. W. *An Introduction to Genetics.* Philadelphia and London; Saunders and Co., 1939

[2] Bridges, C. B. *Genetics*, 1916, **1**, 1 and 107

[3] Beadle, G. W. and Tatum, E. L. *Proc. nat. Acad. Sci., Wash.*, 1941, **27**, 499

[4] Avery, O. T., MacLeod, C. M. and McCarty, M. *J. exp. Med.*, 1944, **79**, 137

[5] Jones, W. *Nucleic Acids.* London; Longmans, Green and Co., 1914

[6] Levene, P. A. and Bass, L. W. *Nucleic Acids.* New York; Chemical Catalog Co., 1931

[7] Brown, D. M. and Todd, A. R. In *The Nucleic Acids* (Ed. Chargaff and Davidson), Vol. 1, p. 409. New York; Academic Press, 1955

[8] Bendich, A. In *The Nucleic Acids* (Ed. Chargaff and Davidson); Vol. 1 p. 81. New York; Academic Press, 1955

CHAPTER 2

THE IMPORTANCE OF NUCLEIC ACIDS IN HEREDITY—INDIRECT EVIDENCE

Evidence of an indirect and direct nature has been obtained concerning the role of DNA in heredity, and it is now widely accepted that it is the genetic material, except in certain viruses which do not contain DNA. In such viruses the genetic information is contained in ribonucleic acid (RNA).

Indirect evidence concerning the role of DNA in inheritance is based on a correlation between the properties shown by DNA and those expected of the genetic material. If genes are composed of DNA the distribution of DNA should parallel that of the genes both in location and in quantity. In higher organisms, chromosomes, which have been shown by genetic and cytological experiments to contain the genes, are composed largely of DNA, protein and RNA. Fish sperm heads which are also known to carry genes contain only DNA and protein, the protein being markedly different from that present in other chromosomes. Thus, DNA is the common component in these two structures which are both known to contain genes. In higher organisms DNA is found only in the chromosomes. There are a few exceptions to this such as *Paramecium aurelia*, certain strains of which have particles containing DNA in their cytoplasm[1]. These so-called *kappa* particles are probably best regarded as self-determining invaders and not as part of the normal cell. It has been suggested that the eggs of certain animals, for example sea urchins, contain DNA in their cytoplasm but this has not been definitely established and its significance is doubtful.

A close correlation between DNA and the genetic material was made possible by work on the giant salivary gland chromosomes which are present in some of the Diptera including Drosophila[2]. In some mutant strains of *D. melanogaster* it has been shown that associated with the mutant character is the loss of a deoxyribonucleoprotein segment from the salivary gland chromosome[3].

The first demonstration that the DNA content of cells remained fairly constant within a species was given by Vendrely and Vendrely[4] who isolated nuclei from various bovine organs and estimated the amount of DNA per nucleus. The amount was very nearly the same for all organs tested (Table 2.1). This amount of DNA per diploid somatic nucleus was also approximately twice that present in the nucleus of a bull sperm, which is haploid. A similar relationship has

TABLE 2.1

Amount of DNA per Nucleus in Beef Tissues[4]

Organ	pg*DNA*	Ploidy
Thymus	6·6	Diploid
Liver	6·4	Diploid
Pancreas	6·9	Diploid
Kidney	5·9	Diploid
Sperm	3·3	Haploid

pg = picogram = 10^{-12} g

since been found for other species. It has also been shown that neither RNA nor nuclear protein has the distribution expected of the genetic material.

An analysis of the distribution of DNA in a polyploid series of yeasts[5] showed that in a series of cultures from haploid to tetraploid

TABLE 2.2

Amount of DNA Phosphorus per Cell in a
Polyploid Series of Yeasts[5]

Ploidy	Amount of DNA phosphorus per cell ($\mu g \times 10^9$)
Haploid	2·26 ± 0·23
Diploid	4·57 ± 0·60
Triploid	6·18 ± 0·54
Tetraploid	9·42 ± 1·77

the DNA content per cell increased in proportion to the number of chromosome sets, that is, with ploidy (Table 2.2). On the basis of these results a genetic role for DNA is a reasonable assumption, but not a necessary one. Deviations from this rule occur only when there is reduplication of the chromosomes without subsequent separation, as in the giant salivary gland chromosomes of Drosophila, and during the process of pollen formation. It should be pointed out that more recently it has been shown that many characters other than DNA content show a ploidy dependence[6].

Once formed, DNA is metabolically very stable and does not take part in cellular metabolism to the same extent as do other cellular constituents. Early evidence for this was obtained from the demonstration that no appreciable incorporation of radioactive phosphorus, adenine or formate into the DNA of non-dividing liver cells occurred in either the rabbit or the rat[7]. Other evidence of metabolic stability has been obtained from experiments in which growing cells are allowed to incorporate labelled precursors into their DNA and are then transferred to unlabelled medium where growth can continue[8]. During growth in this unlabelled medium no loss of the isotopes already incorporated into the DNA was detected. Results obtained by studying the incorporation of labelled precursors into the DNA of regenerating rat liver also showed that the isotopes, once incorporated, were not lost from the DNA. It has been suggested that such stability is a desirable, although not essential, characteristic of the genetic material.

DNA isolated from any source is a mixture of a large number of molecular species (Chapter 4). Nevertheless, if all cells of an organism contain the same genes consisting of DNA it would be expected that the ratio of the four purine and pyrimidine bases present in DNA would be constant for all tissues of the same organism. No significant differences have been found among DNA samples obtained from different organs of the same organism. On the other hand, the base composition of DNA from different species does vary and is a characteristic of the species concerned[9]. The evidence is presented more fully in Chapter 4.

Further evidence for the genetic role of DNA has been obtained from experiments with mutagenic agents. The efficacy of ultra-violet light in producing mutations is maximal at the wavelength (*ca.* 260 mμ) of maximum absorption by DNA. However, the actual wavelength of maximum efficiency is not the same for all systems and light of wavelength close to that of the absorption maximum of proteins is also mutagenic. In addition, it has been shown in other species that energy absorbed by one molecule may be transferred to another before producing any effects, so that absorption of energy by the nucleic acid molecule might be only the first step leading to mutation at quite a different site.

Early experiments with mutagenic chemicals such as sulphur mustard were also inconclusive since it was shown that while, in general, the mustards reacted more readily with DNA than with proteins, some proteins were particularly reactive with mustards[11]. The use of base analogues has however provided more direct evidence indicating a genetic function for DNA. For example, when 5-bromouracil, a thymine analogue, is incorporated into the DNA of cells or virus particles, the mutation rate increases considerably[12].

REFERENCES

[1] Preer, J. R. Jr. *Genetics*, 1948, **33**, 625

[2] Painter, T. S. *Science*, 1933, **78**, 585

[3] Swanson, C. P. *Cytology and Cytogenetics*. London; Macmillan and Co. Ltd., 1958

[4] Vendrely, R. and Vendrely, C. *Experientia*, 1948, **4**, 434

[5] Ogur, M., Minkler, S., Lindegren, G. and Lindegren, C. C., *Arch. Biochem.*, 1952, **40**, 175

[6] Ogur, M. *J. Bact.*, 1955, **69**, 159

[7] Smellie, R. M. S. In *The Nucleic Acids* (Ed. Chargaff and Davidson), Vol. 2, p. 393, New York; Academic Press, 1955

[8] Thomson, R. Y., Paul, J. and Davidson, J. N. *Biochem. J.*, 1958, **69**, 553

[9] Chargaff, E. In *The Nucleic Acids* (Ed. Chargaff and Davidson), Vol. 1, p. 307. New York; Academic Press, 1955

[10] Hollaender, A. and Zelle, M. R. In *First International Photobiological Congress*, p. 148. 1954

[11] Needham, D. M. *Symp. Biochem. Soc.*, 1948, **2**, 16

[12] Zamenhoff, S., de Giovanni, R. and Greer, S. *Nature, Lond.*, 1958, **181**, 827; Freese, E. *Proc. nat. Acad. Sci., Wash.*, 1959, **45**, 622

CHAPTER 3

THE IMPORTANCE OF NUCLEIC ACIDS IN HEREDITY—DIRECT EVIDENCE

The direct evidence that nucleic acids are the genetic material is based mainly on the results of experiments with micro-organisms.

TRANSFORMATION

Transformation was discovered by Griffith[1] in 1928 while he was studying the bacterium *Diplococcus pneumoniae*, which causes pneumonia. Normally the pneumococcal cells have a polysaccharide coat which confers virulence on the organism and gives colonies a smooth, glistening appearance on agar medium. Various types of Pneumococcus may be distinguished on the basis of serological reactions which distinguish between chemical differences in the polysaccharide capsule of the different strains. Serological type specificity is inherited and maintained during repeated subculturing of any strain. The virulent smooth (S) strains infrequently give rise to rough (R) variants which lack the polysaccharide capsule and also the virulence of the parental S strain. Spontaneous mutation from one smooth type to another smooth type or from rough to smooth has never been observed.

Griffith, in his experiments, inoculated mice subcutaneously with large quantities of heat-killed virulent cells of serotype III and small quantities of living rough cells, that is avirulent cells, derived from smooth cells of serotype II. Control groups of mice were inoculated with only one of the strains. Infection did not develop in any of the mice in the control groups. Some of the mice which were inoculated with cells of both the living avirulent and the heat-killed virulent strains did, however, become infected and from these animals living virulent S organisms were isolated. The serotype of the virulent cells isolated from infected animals was III, that is the same as that of the

$$S II \xrightarrow[\text{rough-type cells isolated}]{\text{spontaneously occurring}} R II \xrightarrow[\text{+heat-killed S III cells}]{\text{in mice}} \text{live S III type cells}$$

heat-killed cells used in the experiments. Griffith suggested that the properties of virulence and capsule type had been transferred from

11

the dead S cells to the living R strain and that the latter had been transformed. The progeny of these transformed bacteria also possessed capsules and this encapsulation persisted through successive generations indicating that a hereditary determinant was involved.

Griffith's results were confirmed by a number of other investigators but the initial attempts to demonstrate transformation in an *in vitro* system failed. *In vitro* transformation was firmly established by Alloway[2] who used heated and filtered extracts of S pneumococci to induce transformation in R cells. Rigid controls excluded the possibility of living S forms being present in the extracts. He concluded that one or more constituents of the extract 'supply an activating stimulus of a specific nature in that R pneumococci acquire the capacity of elaborating the capsular material peculiar to the organism extracted'. He also showed that purified capsular material was ineffective in causing transformation but he did not identify the component of the cell extract which was responsible for transformation.

Another decade of intensive research followed before Avery, MacLeod and McCarty[3] in 1944 made the decisive advance by showing that in an *in vitro* system DNA isolated from a smooth Pneumococcus strain would transform rough cells into smooth cells with the same serotype as the cells from which the DNA was derived (*Figure 3.1*). The finding that the transforming principle was DNA was surprising because at that time it was widely believed that only proteins were sufficiently complex to form the genetic material.

The criteria used to establish the nature of the transforming principle were extensive. Elementary analysis of the principle corresponded with that of DNA and chemical and physical methods failed to show the presence of other substances in appreciable quantities. Serological tests could not detect any immunologically active polysaccharides in the preparation. Proteolytic enzymes had no effect on the transforming activity of the preparation but this activity was destroyed by deoxyribonuclease. These criteria, which were used to establish that the transforming principle was DNA, were criticized[4] on the grounds that the chemical methods were not sufficiently sensitive to detect the presence of a small amount of impurity, which might be responsible for the activity. In addition, it was suggested that the evidence obtained by using deoxyribonuclease and some proteolytic enzymes was not conclusive.

To meet these criticisms more highly purified transforming principles were prepared and Hotchkiss[5] reported a preparation of DNA from *Haemophilus influenzae* containing less than 0·02 per cent protein. It has been calculated that about 10^{-14} g of DNA is required to transform a single cell of Haemophilus[4]. Assuming a protein content of 0·01 per cent in such preparations, this amount of DNA corresponds to less than one molecule of protein of molecular weight 10^6. Thus on purely analytical grounds it is possible to eliminate the possibility that protein has a specific role in transformation. There is now little doubt that the active component of the transforming system is DNA and that DNA has been shown thereby to have a specific genetic function.

When rough cells are transformed to smooth cells the capsule of the

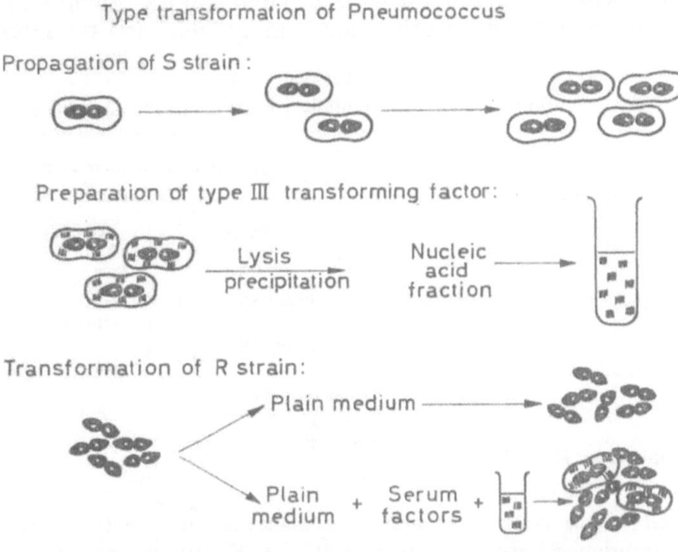

Figure 3.1. Transformation of capsular type in Pneumococcus
(Reproduced by courtesy of W. B. Saunders Co. Ltd.)

newly transformed cells is always composed of the same specific polysaccharide, as determined serologically, as that of the strain from which the DNA was obtained. These transformed cells can in fact be used as a source of DNA for transforming other rough cells to the same smooth type. From this it seems that the transforming principle has two properties characteristic of genes, namely, determination of a specific inheritable property and self-reproduction.

The system used for transformation in Pneumococcus is fairly complex and the frequency of cells transformed for capsule type is usually about one per million. Other simpler systems are now available and with *Bacillus subtilis*[6] and *Haemophilus influenzae* transformation frequencies of one per hundred to one per thousand of the cells treated can be attained. Transformation is no longer restricted to capsule type and a considerable amount of information is now available about the transformation of resistance to antibiotics and the ability to make both biosynthetic and fermentative enzymes[7, 8].

The possibility that the DNA involved in transformation acts as a mutagen and not as genetic material has been virtually disproved by the use of DNA labelled with radioactive phosphorus (^{32}P). In experiments with ^{32}P-labelled DNA it has been shown that the DNA is actually incorporated into the genetic material of the cell presumably by recombination between the transforming DNA and the DNA of

TABLE 3.1

Transformation Involving two Unlinked Markers in Pneumococcus

Donor DNA	Receptor cells	Frequency of transformants	
		Single	Double
pen-r str-r	pen-s str-s	pen-r str-s 0·01 pen-s str-r 0·01	0·0001 pen-r str-r

pen-r = penicillin resistant
str-r = streptomycin resistant
pen-s = penicillin sensitive
str-s = streptomycin sensitive

Data from Hotchkiss[10]

the cell which is transformed[9]. Experiments of this type reveal a correspondence between the amount of DNA incorporated and the number of cells transformed. Although many characteristics may be transformed, usually only one is transformed at a time and it is always a characteristic of the donor strain employed. Thus the mutagen hypothesis of transformation would require the DNA to be a highly specific mutagen corresponding exactly to the characteristics of the donor strain.

The use of drug-resistant strains and strains differing in their biochemical requirements in transformation experiments[7] allowed more precise quantitative experiments to be carried out than had been possible when type specificity was the only marker available. It has been shown[10] that when donor and recipient bacteria differed in several hereditary characteristics, in general each character of the donor strain is acquired independently of the others by the recipient bacteria, that is, the frequency of doubly transformed cells is approximately the product of the frequencies of the two classes of singly transformed cells (Table 3.1). In some experiments, the frequency of double transformants was significantly higher than the product of the frequencies of the single transformants (Table 3.2). Such cases provide evidence that genetic factors can occur linked together in isolated DNA particles. Information obtained in experiments of this type can be used to construct genetic maps by methods analogous to those used with higher organisms. Further evidence that genetic factors may be in the same DNA molecule has been obtained from experiments which

TABLE 3.2

Transformation Involving Two Linked Markers in Pneumococcus

Donor DNA	Receptor cells	Frequency of transformants	
		Single	Double
mtl+ str-*r*	mtl− str-*s*	mtl+ str-*s* 0·01 mtl− str-*r* 0·01	mtl+ str-*r* 0·001–0·002

mtl+	mannitol fermenting	str-*r*	streptomycin resistant
mtl−	mannitol non-fermenting	str-*s*	streptomycin sensitive

Data from Hotchkiss[10]

show that, when the DNA is heated under carefully controlled conditions, genes which recombination studies show to be closely linked lose their transforming ability at almost the same time[11]. Genes which are thought to be unlinked lose their transforming ability independently.

BACTERIOPHAGE

Bacteriophage are viruses which attack bacteria[12]. Perhaps the most intensively studied group are the T bacteriophages which infect *Escherichia coli* cells: they multiply within the host cell and are released after lysis of the cell. Morphologically the T phages are tadpole shaped (*Figure 3.2*) and consist of a protein coat (about 60 per cent) and a DNA core (about 40 per cent) with a small amount of lipid. The DNA is located in the head of the phage. The tip of the tail contains a specific protein which attaches the phage to special sites on the bacterial cell wall. Once attached to a bacterial cell, the phage injects its nucleic acid into the bacterium and in a short time, if the phage is virulent, the bacterium lyses and a large number of phage particles are released. Genetically the released particles are identical with the parent particle.

The role of the phage nucleic acid in viral reproduction was established initially by the work of Hershey and Chase[13]. Phage whose DNA contains radioactive phosphorus can be produced by infecting bacteria which have been grown in medium containing ^{32}P. Phage proteins contain no phosphorus hence only the DNA is labelled. Bacteria which have been grown in medium containing radioactive sulphur (as sulphate) can similarly be used to produce phage whose proteins contain ^{35}S. By this means it is possible to distinguish easily

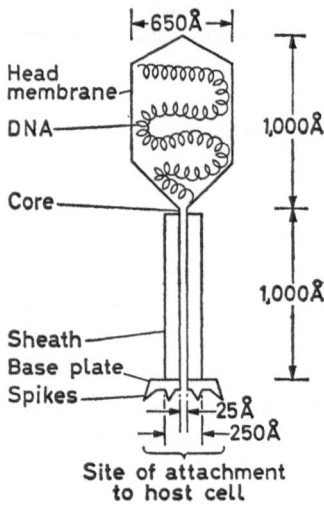

Figure 3.2.

between phage protein and phage DNA after bacteria have been infected by labelled phage. In their experiments Hershey and Chase infected bacteria either with phage whose protein was labelled or with phage whose nucleic acid was labelled. A few minutes after infection the complexes of phage and bacterium were separated by vigorous agitation in a blender and the bacterial cells were examined for radioactivity. Agitation in the blender separated the phage coat from the bacterium but did not prevent the production of progeny phage in the infected bacterial cells. When ^{32}P-labelled phage was used it was found that virtually all of the ^{32}P entered the bacterial cell and that a considerable portion of the ^{32}P could be recovered in the progeny phage.

With ^{35}S-labelled phage, however, following treatment in the blender, less than 3 per cent of the radioactivity was associated with the cells and none of this appeared in the progeny phage. If this small percentage of the protein which may enter the cell is ignored, then it must be concluded that the inherited characteristics of the virus were transmitted through the nucleic acid portion and that the protein plays no direct part in heredity. Support for this conclusion has been obtained from experiments in which phage labelled with ^{32}P were stored at low temperatures[14]. Under these conditions the phage were progressively inactivated as a function of the number of atomic disintegrations which had occurred. ^{32}P-labelled phage have also been used in crossing experiments which showed that the radioactivity was associated with the genetic material of the phage[15].

TOBACCO MOSAIC VIRUS

While all viruses studied so far contain nucleic acid, not all of them contain DNA. Most viruses which attack higher plants contain RNA as their nucleic acid component. Tobacco mosaic virus (TMV)[16] is a virus of this type. The TMV particle consists essentially of a single RNA chain surrounded by identical small protein molecules arranged in a helix (see page 69; Figure 6.2; Plate 2).

Several methods are known to degrade the virus particles and separate the protein from the nucleic acid. Schramm[17] and his co-workers obtained viral protein which, in the absence of RNA, re-aggregated to form rods which appeared very similar to TMV particles. These virus-like particles, however, were not infective. Similar results were obtained with turnip yellow mosaic virus in which it is also possible to obtain virus-like particles which contain protein but no RNA[18] (see Chapter 6). This evidence suggested that the protein of the viruses was not responsible for viral infectivity. Fraenkel-Conrat and Williams[19] showed that if the protein was allowed to reaggregate in the presence of RNA isolated from the virus, apparently normal virus particles, containing both RNA and protein, were formed in about 50 per cent yield. These reconstituted particles were infectious and indistinguishable from natural TMV of the same strain.

While viral protein alone was not infectious, viral RNA was able to infect plants but with a very low efficiency due to the ready inactivation of the RNA during the assay procedure. Plants infected with RNA by itself produced new virus particles complete with protein coats, suggesting that all of the genetic information present in the virus was carried in the nucleic acid component. Confirmation of this was obtained from an ingenious reconstitution experiment[19] in which the RNA and protein of two different strains of virus were separated. Particles were then reconstituted in which the nucleic acid of one strain was recombined with the protein of the other. The symptoms of the disease resulting from these 'hybrid' viruses are always similar to those of the strain which supplies the nucleic acid component of the reconstituted virus. Additionally if the 'hybrid' virus is allowed to replicate and progeny virus is then isolated, the protein of the new virus particles is found to be identical with that supplying the RNA component. The RNA of a given virus strain thus causes the production of more virus of the same strain and therefore must carry all the genetic information necessary for its production. Confirmation of this viewpoint has been obtained from experiments with a number of other animal and plant viruses in which it has been shown that all of the genetic information necessary for virus replication is carried by the ribonucleic acid component[20].

SEXUAL REPRODUCTION IN BACTERIA

In the bacterium *Escherichia coli* sexual reproduction[21] occurs in a process in which two conjugating cells are united by a tube through which the genetic material of the donor cell is transferred to the recipient cell. It has been shown that the transfer of genetic information during conjugation is a relatively slow process and may be interrupted by separation of the conjugating cells[22]. By separating cells at various times after 'mating' has begun, it is possible to determine the time at which a given marker is transferred from the donor to the recipient cell. The order of entry of the genes is the same as the sequence deduced from chromosome mapping experiments. Using donor cells labelled with ^{32}P it is possible to show that the DNA enters the recipient cells at the same rate as the genes. The amount of ^{32}P transferred correlates fairly well with the fraction of all the genes which has entered the recipient cell.

Experiments in which donor bacteria were labelled with ^{32}P and stored at low temperature, either before or after mating, have also given evidence that DNA is the genetic material[23]. In experiments in which radioactive bacteria are mated with unlabelled bacteria before being stored frozen, it has been found that the number of recombinants obtained decreases with time, that is, with ^{32}P decay. If the labelled cells are stored frozen before mating takes place the size of the piece of chromosome transferred to the recipient cell decreases with time. Both types of experiment show that phosphorus plays a key structural role in the genetic material which is compatible with nucleic acid being the material of which genes are composed.

More direct evidence for the transfer of DNA from donor to recipient during conjugation has been obtained in experiments in which the DNA of the donor bacteria was labelled with tritiated thymidine[24]. Since thymidine is incorporated only into DNA (Chapter 4), the amount of DNA transferred could be measured and correlated closely with the amount of genetic material transferred.

HIGHER ORGANISMS

To determine whether transformations similar to those induced in micro-organisms could be induced in vertebrates, DNA from male Khaki Campbell ducks has been injected into Pekin ducklings with the surprising result that the characteristics of some of the treated ducklings were altered[25]. This result was surprising since the modifications were expected to appear only in the progeny of the treated ducklings. The same authors have since reported that the newly acquired characteristics have been transmitted from the modified animals to the third generation progeny[26]. Unfortunately they have been unable to repeat their original experiment, so that the significance of the results is, for the present, doubtful.

REFERENCES

[1] Griffith, F. *J. Hyg.*, *Camb.*, 1928, **27**, 113
[2] Alloway, J. L. *J. exp. Med.*, 1932, **55**, 91
[3] Avery, O. T., MacLeod, C. M. and McCarty, M. *J. exp. Med.*, 1944, **79**, 137
[4] Zamenhof, S. In *The Chemical Basis of Heredity* (Ed. McElroy and Glass), p. 351. Baltimore; Johns Hopkins Press, 1957
[5] Hotchkiss, R. D. In *Phosphorus Metabolism* (Ed. McElroy and Glass), Baltimore; Johns Hopkins Press, 1962
[6] Spizizen, J. *Proc. nat. Acad. Sci.*, *Wash.*, 1957, **43**, 694
[7] Goodgal, S. H. and Herriott, R. M. In *The Chemical Basis of Heredity* (Ed. McElroy and Glass), p. 336. Baltimore; Johns Hopkins Press, 1957
[8] Spizizen, J. *Proc. nat. Acad. Sci.*, *Wash.*, 1958, **44**, 1072
[9] Gareñ, A. and Skaar, P. D. *Biochim. biophys. Acta*, 1958, **27**, 457
[10] Hotchkiss, R. D. *J. cell. comp. Physiol.*, 1954, **45**, Suppl. 2, 1
[11] Roger, M. and Hotchkiss, R. D. *Proc. nat. Acad. Sci.*, *Wash.*, 1961, **47**, 653
[12] Boyd, J. S. K. *Biol. Rev.*, 1956, **31**, 71
[13] Hershey, A. D. and Chase, M. *J. gen. Physiol.*, 1952, **36**, 39
[14] Stent, G. S. *Cold Spr. Harb. Symp. quant. Biol.*, 1953, **18**, 255
[15] Stent, G. S., Fuerst, C. R. and Jacob, F. *C. R. Acad. Sci.*, *Paris*, 1957, **244**, 1840

[16] Gierer, A. *Progr. in Biophys.*, 1960, **10**, 299;
Fraenkel-Conrat, H. *Design and Function at the Threshold of Life: The Viruses.* New York and London; Academic Press, 1962
[17] Schramm, G. *Ann. Rev. Biochem.*, 1958, **27**, 101
[18] Markham, R. *Disc. Faraday Soc.*, 1951, **11**, 221
[19] Fraenkel-Conrat, H. and Williams, R. C. *Proc. nat. Acad. Sci., Wash.*, 1955, **41**, 690
[20] Colter, B. B., Bird, H. H., Moyer, A. W. and Brown, R. A. *Virology*, 1957, **4**, 522
[21] Hayes, W., Jacob, F. and Wollman, E. L. In *Methodology in Basic Genetics.* (Ed. Burdette), p. 129. San Francisco; Holden-Day, Inc., 1963
[22] Wollman, E. L. and Jacob, F. *C. R. Acad. Sci., Paris*, 1955, **240**, 2449
[23] Jacob, F. and Wollman, E. L. *Symp. Soc. exp. Biol.*, 1958, **11**, 75
[24] Wollman, E. L. and Stent, G. S. Quoted in *Sexuality and the Genetics of Bacteria*, by Jacob, F. and Wollman, E. L. New York; Academic Press, 1961
[25] Benoit, J., Leroy, P., Vendrely, C. and Vendrely, R. *C. R. Acad. Sci., Paris*, 1957, **244**, 2320
[26] Benoit, J., Leroy, R., Vendrely, R. and Vendrely, C. *Trans. N.Y. Acad. Sci.*, 1960, **22**, 494

PART 2

THE MOLECULAR AND STRUCTURAL BASIS

CHAPTER 4

DEOXYRIBONUCLEIC ACID

LOCATION

The early history of the nucleic acids, starting in 1868 with their initial isolation by Miescher from pus cells and then from salmon sperm heads, to the recognition and proof of the nature of their chemical constituents, has been described in Chapter 1. By 1930 two types of nucleic acid had been identified and differentiated by the sugar component present in each, namely, deoxyribose in DNA and ribose in RNA. For some time it was thought that DNA was characteristic of animal cells and RNA of plant cells. This distinction was shown to be false, for the presence of DNA has been demonstrated in a very wide variety of cells and it is now generally agreed that, with the possible exception of bacteria whose nuclear apparatus is often less clearly distinguished, it is almost completely confined to the nucleus, where it resides in the 'chromatin'—the component of the chromosomes which binds basic, cationic dyes. That it is DNA and not RNA which is responsible for the staining properties of chromosomes has been amply demonstrated by means of cytochemical tests specific for DNA. Thus the Feulgen nucleal reaction depends on the greater ease, relative to RNA, with which mild acid treatment can release the purines from DNA and so expose *aldehydo*-groups, the presence of which can then be revealed by Schiff's reagent. The dependence of DNA content on polyploidy has been discussed in Chapter 2 (page 8).

CHEMICAL STRUCTURE

The component units of DNA have been deduced from extensive studies of chemical hydrolysis and enzymic degradation[1-3]. Acid hydrolysis yields the purines adenine (I) and guanine (II) and the deoxyribodiphosphoric acids of thymine and of cytosine (thymidine-3',5'-diphosphate and deoxycytidine-3',5'-diphosphate); under more drastic conditions the free pyrimidines, cytosine (III) and thymine (IV), can be obtained. Thymine is a constituent of DNA whereas the unsubstituted form, uracil, is absent, apparently because there is no mechanism, in *Escherichia coli* at least, for phosphorylating deoxyuridylate to the triphosphate form necessary for polymerization to the polydeoxyribonucleotide (*see* page 38). Intestinal phosphatases act on DNA to give the four monodeoxyribonucleotides of the 'bases'* I–IV in which the phosphoryl group is attached to the 5'

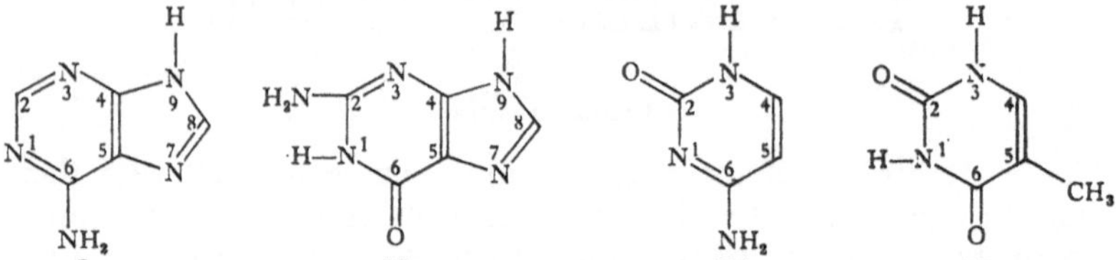

position of 2-deoxy-D-ribose (V) and the enzyme which is specific to DNA, deoxyribonuclease, degrades it to a complex mixture of oligo-nucleotides and di-nucleotides. From such evidence it has been deduced that the structural elements in DNA are related in the following way:

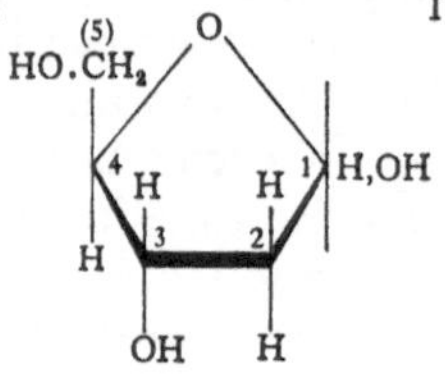

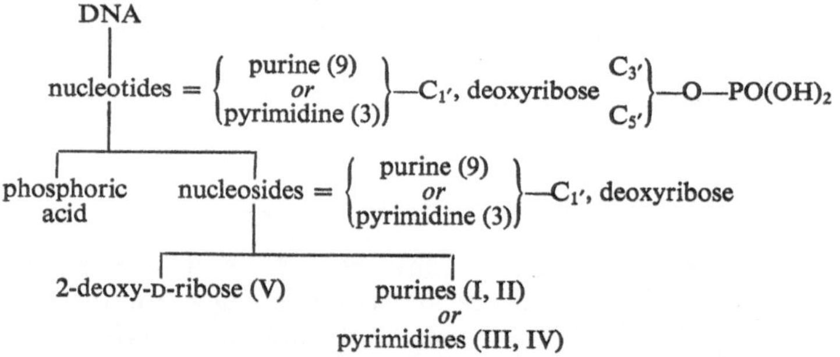

The bases I–IV are the most commonly occurring but there are important exceptions. The DNA of higher plants contains up to 20 per cent of 5-methyl cytosine and it occurs to a lesser degree in

* So-called, because of the proton-accepting character and nitrogen content of the purine and pyrimidine ring structures.

animal DNA. The T-even bacteriophages which attack *E. coli* contain 5-hydroxymethyl cytosine instead of cytosine and glucose joined to many of the hydroxymethyl groups. Various homologues of thymine, such as 5-chloro, 5-bromo and 5-iodouracil can replace this base, and from a thymine-less mutant of *E. coli* a DNA can be isolated with 20 per cent of its thymine replaced by 6-methylamino-adenine[4]. These are exceptions, however, and bases I–IV predominate in the vast majority of DNA. They are linked through the 9 position of the purine ring or the 3 position of the pyrimidine to the C_1 atom of the deoxyribose ring (V), so this linkage is glycosidic and is of the β-type.

The base-sugar units are termed nucleosides and phosphorylation of the sugar ring in these substances gives a nucleotide. In deoxyribose two positions, at $C_{3'}$ and $C_{5'}$ (the prime is used to denote positions in the sugar ring as distinct from the base rings), are available for this phosphoryl group, and this gives rise to two distinct series of nucleotides. The degradative evidence already mentioned, careful comparison of the degradation products with synthetic deoxyribonucleotides[2] and the titration evidence that phosphodiester bonds constituted the main internucleotide linkage[1,3,5] has led conclusively to the formulation of the main chain in DNA as a linear, unbranched polynucleotide (VI, abbreviated to VII) with a recurring 3', 5'-internucleotide linkage, joining many thousands of nucleotide units.

COMPOSITION

Originally it was thought that the bases were present in equivalent amounts, but chromatographic analysis of a large number of samples of DNA from a variety of sources disproved this[6]. The following relationships have emerged from this work (letters, A, G, C, T denote numbers of moles of the bases, I–IV, respectively).

(1) The sum of the purine nucleotides equals the sum of the pyrimidine nucleotides, that is, $(A + G) = (T + C)$.

(2) Adenine and thymine are present in equal amounts, and so also are guanine and cytosine, that is, $A = T$; $G = C$.

(3) The number of 6-amino groups equals the number of 6-oxo groups, that is, $A + C = G + T$.

(4) The composition of DNA, which represents the average for a mixture of molecules, is characteristic of the species from which it is derived. This composition may be expressed as the mean guanine-cytosine content (GC) defined as $(G + C)/(G + C + A + T)$. For bacterial species the GC content varies[7–9] from 0·25 to 0·75. The range is narrower with higher organisms, namely, *ca* 0·4–0·44 for vertebrate DNA; 0·34–0·44 for invertebrate DNA and for the higher plants; 0·35–0·66 for fungi and algae[7]. Taxonomically related bacteria often have similar GC contents, so much so that it has been suggested

that bacterial species and other micro-organisms might be re-classified by the GC content of their DNA[8, 9].

(5) No significant differences in total composition have been found in DNA from different tissues of the same species.

When 5-methyl cytosine (MC) is present the sum (C+MC) replaces C in the above relationships, and 5-hydroxymethyl cytosine (HMC) completely replaces C in the DNA of the T-even bacteriophages which attack *E. coli*, so that HMC = G. In the DNA containing 5-halogenouracil derivatives these compounds replace the thymine and are equivalent in amount to adenine. The relatively large amounts of 6-methylamino adenine (MAA) which replace thymine in a thymine-less mutant of *E. coli*, when growing in the absence of thymine, constitute a different problem since the total $(T+MAA) \neq A$; the effect is, however, lethal to the cells, so the DNA is probably abnormal. The relationship (2) seems to suggest some kind of pairing of bases within the DNA structure. A possible structural basis for such pairing had already been tentatively proposed to explain the anomalous titration behaviour of DNA[10]. The dissociation curves obtained when solutions of DNA, pH 6–7, were first titrated to pH 2·5 or to pH 12 were irreversible and differed markedly from the reversible curves obtained on back-titration from these pH values to neutrality[5]. Since altering the pH from neutrality ionizes the amino- or the 1,6-NH.CO-groups by addition or removal of a hydrogen ion, respectively, it was suggested that these groups must each be involved in labile linkages, probably hydrogen bonds, which are ruptured on ionization. The exact coincidence of the back-titration curves led to the conclusion that these hydrogen bonds linked the titratable amino and 1,6-NH.CO-groups[5]. The significance of this conclusion from the titration behaviour and of the analytical relationships (1–3) only became fully apparent when they were considered in conjunction with the double-helical structure for DNA which x-ray diffraction studies later suggested.

All the molecules of DNA in any one preparation do not have the same composition. That the preparations are heterogeneous has been demonstrated by chromatographic fractionation: thus, 7 fractions of calf thymus DNA have been separated[11] which vary in $(A+T)/(G+C)$ ratio from 0·99 to 1·79, though each individual fraction contained quantities of individual bases in agreement with the relations (1–3). The preparations of DNA contain molecules which differ in the proportions of the bases in each of them. This is confirmed by the observation that thymus DNA fractions of this type all exhibited the same x-ray patterns as the original DNA, that characteristic of the double-helical structure[12]. But deviations from the relations (1–3) in individual fractions of thymus DNA have been reported and in an average molecule containing 20,000 nucleotides several hundred bases may be unpaired. Not surprisingly, the singly stranded DNA from the bacteriophage ϕX-174 also had a composition which failed to accord with these relations[13].

By making use of the relation between composition and both the observed increase in ultra-violet absorption with heating of DNA solutions and the density of the DNA in a density gradient of caesium chloride, it has been possible to compare more easily the relative heterogeneity of DNA from a number of sources[7, 14]. The distribution of GC content of the DNA molecules of an organism is, with few exceptions, uni-modal and of relatively narrow range. Bacterial DNA have narrower distributions than those of higher organisms, so much so, that, if the mean GC contents of DNA of two bacterial species are different by 10 per cent, there are few DNA molecules of the same GC content common to the two species. Since the metabolism of bacteria shares many common features, this observation raises important questions concerning the universality of the genetic coding relation between DNA and protein (Chapter 12). By studying the density gradient distribution of fragmented DNA it has been shown[7] that the small heterogeneity of the base composition among the DNA molecules of an organism is the same as that in smaller regions of about $\frac{1}{10}$ of the original length: in very much shorter regions, a few nucleotides in length, departures from the norm have been deduced.

CONFIGURATION

The earlier work of Astbury (1947) demonstrated a repeat distance of 3·4 Å, and led him to suggest that the flat aromatic base rings were probably arranged like a 'pile of pennies'. These studies were greatly extended by a group of investigators at King's College, London, who showed that moist fibres of DNA from a variety of sources all give a similar x-ray diffraction pattern, of a form which indicates that they contain helical polynucleotide chains[15]. On this basis and by means of model building, Watson and Crick[16] were able to propose a model which allows for the pairing of bases indicated by the other evidence and could explain how DNA containing different amounts of four bases of different dimensions could be accommodated into a structure of great regularity, whatever the value of the ratio $(A+T)/(G+C)$. They made the important assumption that the number of

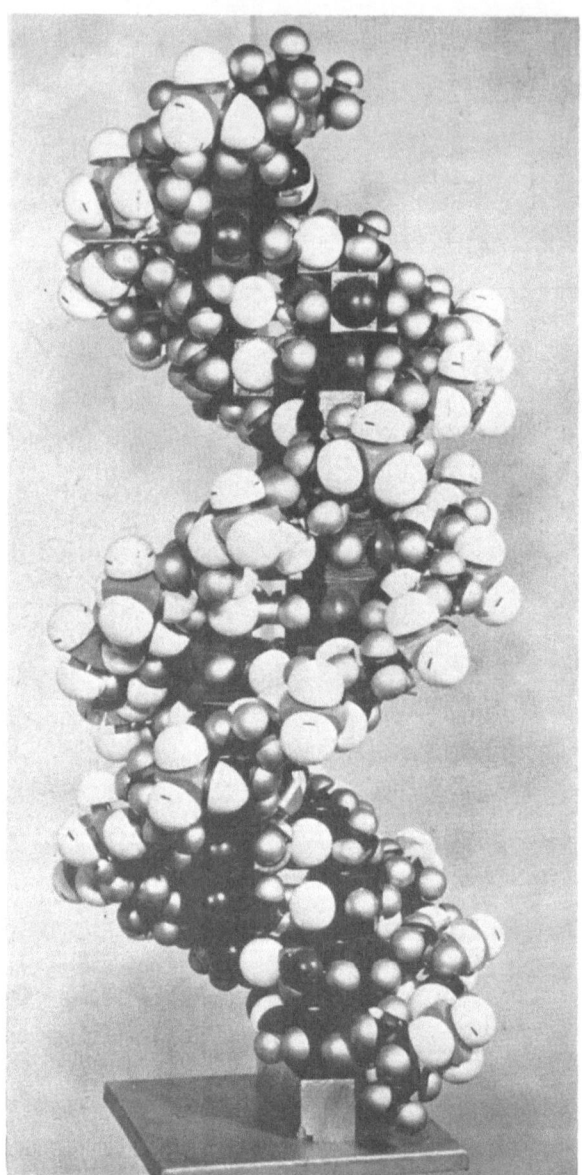

PLATE 1

Model of the paracrystalline B form of DNA showing 16 nucleotide pairs. The white spheres represent oxygen atoms, the grey segments phosphorus atoms, the small silver spheres hydrogen atoms. The black spheres are carbon atoms and the bases within the spirals are denoted by appropriately shaped blocks. This model was constructed at King's College, London

(Reproduced by courtesy of the Editor of *Biological Reviews of the Cambridge Philosophical Society*, and of Prof. M. H. F. Wilkins, Biophysics Dept., King's College, London)

To face p. 29

helical polynucleotide chains in the DNA molecule is two and they found that a structure was possible in which two chains are wound spirally in right-handed helices around the same axis but running in opposite directions (the chain has a direction determined by the sequences of atoms in the $C_{3'}$—O—PO—O—$C_{5'}$ phosphodiester linkage). The sugar–phosphate chains were regarded as held together by pairs of hydrogen bonds, between a purine base attached to one chain and a pyrimidine base attached to the other, the flat rings of these bases lying perpendicularly to the main axis of the helices (VIII). The main features of this structure, at first somewhat hypothetical, have now been proved by careful and elegant x-ray studies not only of the sodium salt but also those of other alkali metals. From this it transpires that in the fibrous solid state DNA has the following configurations[17]:

(*A*) is a very compressed and highly crystalline structure which exists at 75 per cent relative humidity and in which there are 11 base

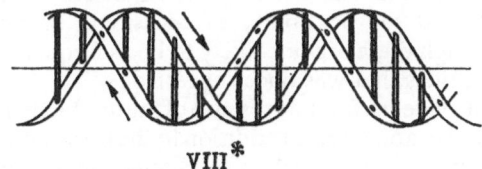

VIII*

pairs inclined at about 70 degrees to the helix axis in each complete turn (28·1 Å along the axis) of the double helix of a diameter 18 Å.

(*B*) is a slightly more extended paracrystalline structure which is formed reversibly from the sodium salt in the *A* form when the relative humidity is increased to about 90 per cent. In this structure there are 10 nucleotide base pairs for every complete turn of the double helix (33·6 Å along the helix axis) and the flat base rings are disposed perpendicularly to the axis: this means that the separation between the base rings along the axis direction is 3·36 Å. The diameter of the double helix is again 18 Å. Plate 1 is a photograph of the model of the *B* structure which has been constructed at King's College, London.

(*C*) has been observed when the relative humidity above the lithium salt of DNA in the *B* configuration is reduced from 66 per cent to 32–44 per cent. The base planes, instead of lying perpendicularly to the helix axis, are slightly tilted and there are 9·3 instead of 10 base

---

* The two ribbons symbolize the two phosphate–sugar chains and the vertical rods the pairs of bases holding the chains together. The horizontal line is the fibre axis.

pairs per turn. The C configuration of the lithium salt of DNA can be obtained, in principle, by twisting the B form to give a structure in which the bases are more tilted and more closely packed.

Direct evidence that the chains run in opposite directions has been obtained independently in studies of 'nearest neighbour' frequencies utilizing labelled nucleoside 5'-triphosphates which are then enzymatically degraded[18]. The same work has also confirmed that the pairing is A–T and G–C.

From the point of view of the behaviour of solutions of DNA, structure B is the most significant, since it is the form in which the nucleic acid exists when combined with protein[19] and is also the form stable in the presence of a large excess of water. There is now considerable agreement between the observed x-ray diffraction intensities and those calculated for DNA in this configuration. The x-ray evidence is that the double-helical structure must hold for lengths of at least 20 turns of the helix, that is, for a molecular weight of the order of 10^5, but it cannot determine whether or not the molecule is regular over a greater length and the possibility of occasional breaks cannot be excluded, although these are not likely to be as frequent as the one in every 50 nucleotides that was once suggested.

Independent evidence for the presence of the double-helical structure in solution has also appeared, in addition to the titration and analytical evidence already discussed[20,21]. The changes in the viscosity and molecular weight of DNA in solution which occur when it is subjected to γ-rays can only be explained if a degradative fracture is produced by two independent breaks induced by the free radicals which the γ-rays produce. Evidence of the same kind for a two-stranded structure has also been obtained from studies of the degradation of DNA in solution by enzymes and of solid DNA by x-rays.

This accords with an earlier deduction from x-ray scattering measurements that the 3·4 Å spacing characteristic of the helical form of the solid state is also present in concentrated DNA gels. Studies of the scattering of x-rays at low angles from dilute solution and gels of DNA show that the DNA behaves like a long, rigid rod of a mass per unit length which is equal to that expected for the double-helical B structure[22]. This structure also affords the best explanation of the irreversible increase in ultra-violet absorption and of other physico-chemical changes which occur when DNA solutions are heated and subjected to extremes of pH and to other agents. These effects are explained in terms of a change in the original configuration of the DNA so that the ordered arrangement of the base rings which absorb the ultra-violet light is destroyed. This process may occur without rupture of covalent bonds and is termed 'denaturation' by analogy with the similar phenomenon in proteins. This denaturation is an example of a transition from a helical to a randomly coiled configuration[20, 21, 23].

A novel feature of the helical structure is that specific base-pairing is postulated which means that the strands are complementary, since the nucleotide sequence on one strand determines that on the other. In the original proposal of Watson and Crick[16], the 1,6-N:C(NH₂)-group of an adenine ring, attached to one helix, was regarded as linked by a pair of hydrogen bonds* to the 1,6-NH.CO-group of thymine attached to the other helix (IX): cytosine and guanine on each of the two helices were regarded as similarly linked (X, omitting the asterisked bond). This system of pairing of the bases in a plane is only one among many possibilities[24, 25], all of which may also exist between disordered ('denatured') polynucleotide chains in solution. Alternative schemes of hydrogen bonding in a double helix have been proposed. Admissible schemes all have to agree with the titration and analytical data but the base-pairing proposed by Watson and Crick (IX, X) is the only one which will satisfactorily fit the dimensions and detailed x-ray diffraction results[17] and infra-red studies of the tautomeric forms of the bases. On the basis of more accurate bond lengths and angles in the bases, it has been proposed that a third hydrogen bond links the guanine 2-amino-group and the cytosine 2-oxo-group[26] (X, asterisk). This seems quite feasible and need not be inconsistent with the titration evidence. It also provides an explanation of the relatively lower stability of the linkage between adenine and thymine compared with that between guanine and cytosine. This was observed first with respect to the action of γ-rays on DNA in solution[27] and later with respect to the denaturing effect of heat[14]. Since molecular orbital calculations[28] indicate a greater stabilization by resonance for X (with 2 hydrogen bonds) than for IX, the presence of a third hydrogen bond in X cannot be surely inferred from its greater stability.

---

* In a 'hydrogen bond', the relatively small hydrogen atom links two other atoms and so constitutes a bond between them. It is denoted by —H ⋯ so that a hydrogen bond between atoms A and B is written A—H ⋯B. Although the energy of formation of this bond is small, about 4–7 kcal/mole, and the bond is therefore weak, it is of very great importance in aqueous and biological systems on account of the large number which can be formed, especially those involving OH groups. The bond is essentially electrostatic and forms most readily between atoms such as O, N, F, which are relatively electronegative with respect to hydrogen. The bond is strongly directional in that its energy is lowest when the hydrogen lies on the line joining the two linked atoms, not necessarily at the mid-point; the arrangement is, however, not always exactly linear. Atoms so linked must be no more than 3·4 Å apart.

The double-helical structure permits not only the maximum number of intramolecular hydrogen bonds to be formed but also the maximum interaction between the aromatic base rings of successive nucleotide units. For this reason, and because the double-helical structure of DNA is relatively less stable in non-aqueous solvents, the importance of so-called 'hydrophobic bonds' in maintaining the DNA structure has subsequently been stressed[29]. This type of interaction, which is also thought to be important in protein structure, is the apparent attraction for each other of non-polar residues in a macromolecule, and is a consequence of the tendency of a polar solvent, such as water, to maximize the number of strong solvent–solvent interactions. From this point of view, the DNA structure may be regarded as an unusually elongated detergent micelle[29].

One of the most striking features of the x-ray work has been that DNA from a variety of sources gave the same diffraction pattern whether from stored genetic material (sperm, bacteriophage, wheat germ), from rapidly dividing cells associated with rapid DNA turnover and protein synthesis (acute leukemic leucocytes, developing granulocytes, mouse sarcoma, bacteria in the logarithmic phase), from more slowly dividing tissues with considerably protein synthesis (liver, calf thymus, lymphatic tissue) or from metabolically 'inert' nuclei, such as those of chicken erythrocytes. Moreover, the same characteristic features can be observed in sperm heads, isolated nucleoprotamine, artificial complexes of DNA with protamine and isolated sperm DNA, which demonstrates that the structures observed in the isolated DNA are not artefacts and exist in the intact nucleoprotein in the cell[15, 20]. The similarity of the x-ray patterns of fractionated DNA (*see below*) to that of the original DNA indicates that the DNA does not exist in significant amounts in forms other than the double helix[30].

Evidence has been presented that DNA in cells, in which it is being rapidly synthesized, are linked laterally in pairs, and that separate double-helical molecules only exist when the DNA is not replicating. A scheme of replication has been based on this proposal[31] (*see* Chapter 8). Nevertheless, the masses per unit length deduced from the low angle x-ray scattering of solutions of DNA[22] and from autoradiographs of λ-phage labelled with tritium[32] (*see* page 35) do not indicate any tendency for DNA to associate in parallel bundles. Moreover, the proposal does not accord with the similarity, already mentioned, between the x-ray diffraction patterns of DNA from rapidly proliferating cells and of non-replicating DNA. The relation between these various observations is not yet clear.

It is noteworthy that, from the bacteriophage φX-174, DNA has been isolated which has none of the behaviour associated with the double-helical form and is believed to be singly stranded DNA[33]. Only one molecule of this DNA is obtained from each phage particle. Its existence implies either that a distinct mode of DNA replication occurs in this bacteriophage or that singly stranded DNA is more generally operative in other duplicating systems than had been

3

previously realized. In either case it appears that the essential genetic information in DNA can be carried by a single strand.

MOLECULAR SIZE

The determination of the molecular weight of a charged macro-molecule such as DNA presents many theoretical and practical problems. The two most widely used methods at present are a combination of sedimentation coefficient at infinite dilution with intrinsic viscosity, employing appropriate hydrodynamic equations, and the method of light scattering. Electron microscopy has confirmed a diameter of 20 Å for DNA molecules but cannot readily determine the length. The respective merits of these methods have been much discussed and have been surveyed elsewhere[3, 20, 21]. For the present purposes it may be said that the molecular weights of most preparations of DNA come within the range $5-12 \times 10^6$, with the majority of preparations having a molecular weight of about 6×10^6. This implies molecules containing 9,000 nucleotides, extending over a total length along the helical axis of

$$3 \cdot 36 \text{ Å} \times (6 \times 10^6)/(2 \times 330) \simeq 30,000 \text{ Å} (3 \mu)$$

assuming the 3·36 Å internucleotide spacing of the B structure (330 = average molecular weight of a nucleotide residue). In solution the molecule has a radius of gyration (the mean radius about its centre of gravity) of about 1500–2500 Å and a mean end-to-end length of the order of 5000 Å or more, which shows that it must be coiled. The hydrodynamic and light-scattering measurements both indicate that in solution its configuration is that of a stiffened coil, rather than that of a rigid rod or of a completely random coil. The length of the helical DNA molecule (3μ) is sometimes greater than the largest dimension of the structure from which it has been isolated (for example the head of a bacteriophage of diameter 800 Å $= 0 \cdot 08 \mu$) so that some sort of 'super-folding' must occur *in vivo* (*see* page 67, Chapter 6).

DNA preparations from several sources have been shown to consist of a mixture of molecules of various molecular weights, so that the values already quoted are only averages. This heterogeneity with respect to molecular weight had earlier been suspected from the shape of the light-scattering envelopes and has now been substantiated by direct observation in the electron microscope, when the DNA molecules are revealed as long thread-like structures, and by measurement of the distribution of sedimentation coefficients, which can be correlated with the distribution of molecular weight[3, 20, 21]. The evidence points to the distribution of size being that produced by breaking at random a molecule very much longer than the observed mean. This raised the question whether the DNA as isolated really represented the heterogeneity of size of DNA in the original cells, or whether it was the result of random rupture during extraction of larger and more uniform molecules originally present. Certainly the DNA of highest molecular weights and least heterogeneity were those obtained from bacteriophage, for which the likelihood of breakdown during extraction was least. Thus, by an independent and ingenious technique based on direct counting of the radioactive atoms in ^{32}P-labelled DNA it appeared to be proved[34] that the DNA in the T2 bacteriophage which attacks E. coli contained one large piece of molecular weight about 45×10^6, comprising 40 per cent of all the DNA, and smaller pieces of weight about 12×10^6. However even the large weight of 45×10^6 for the DNA of this bacteriophage soon transpired to be only a lower limit when it was found[35] that the DNA after its extraction is extremely susceptible to degradation by shearing forces, even those arising when solutions are forced through a hypodermic needle to fill an ultra-centrifuge cell, and even more when solutions are agitated in blenders. When such degradative procedures were avoided, the DNA obtained from T2 bacteriophage was shown[36, 37] by autoradiographic methods with ^{32}P to be one physical entity in each bacteriophage and to have a molecular weight between 90×10^6 and 150×10^6, probably at least 130×10^6.

The DNA of T5 and λ E. coli bacteriophages have been shown[37] by sedimentation studies to have molecular weights of 81×10^6 and 46×10^6, respectively, by using the relation between sedimentation coefficient and autoradiographic molecular weight deduced for T2 bacteriophage[36]. Such molecular weights correspond to the entire DNA in each bacteriophage, which must therefore contain only one molecule. Autoradiography of a preparation of the DNA of λ bacteriophage which contained tritiated-thymine, showed[32] that it had a maximum length of 23 μ, and therefore a mass to length ratio of about 2×10^6 mol. wt. units/μ, which is the value expected for the B configuration of DNA. The DNA of λ bacteriophage is therefore all in one double helix.

At the other end of the scale, the small bacteriophage ϕX-174 also contains only one DNA molecule with a molecular weight of $1 \cdot 7 \times 10^6$, but this is singly stranded, and, surprisingly, appears to be in the form of a circle[33]. Thus it is probable that all bacteriophage contain only a single molecule of DNA and this has naturally led to the suggestion that there may be only one molecule of DNA in each chromosome, even of larger organisms[35]. If this were true for *E. coli*, the DNA content of this organism implies that the DNA would have a molecular weight of about 4×10^9. This now seems more likely than could have been suspected a few years ago, for autoradiography of DNA extracted under very mild conditions from *E. coli* labelled with tritiated thymine shows some DNA molecules to be as much as 400 μ long, indicating a molecular weight of 10^9 or more[38]. Moreover, electron micrographs of protoplasts of *M. lysodeikticus* reveal very long continuous threads of deoxyribonucleoprotein[39]. Thus both procedures support the suggestion that all the DNA in a bacterium may exist as a single molecule. Even so, in the larger chromosomes of higher organisms, it still seems likely that several DNA molecules could be linked together in the chromosome by small amounts of protein, possibly involving metal ions, but further knowledge of the structure of nucleoproteins and of chromosomes is required to substantiate this view (*see* Chapters 6 and 7).

SEQUENCE OF NUCLEOTIDES

As will be discussed later, the sequence of nucleotides in DNA controls the structure of proteins, through a RNA intermediate (Chapter 11). Not surprisingly it has been very difficult to obtain direct chemical information concerning sequences of nearly 10^4 bases in each of the two chains. The chemical methods[40] employed have mostly utilized the greater lability to acid of the purine to sugar linkage compared with the pyrimidine to sugar linkage. There seems to be general agreement that sequences of pyrimidines of three and over occur more frequently in calf thymus DNA, though not in some other DNA, than would be expected for a random distribution. The sequence purine-(C_2T)-purine is less frequent than in a random sequence; there is a slight tendency in calf thymus DNA for a purine nucleotide to be followed by another; and certain sequences of two pyrimidines occur at more than random frequency. When 5-methyl cytosine replaces cytosine it does so preferentially next to guanine.

An elegant method has been devised[18] which utilizes the DNA polymerase system described on page 38. The DNA being studied is used as the primer for the synthesis of new polynucleotide from mixtures of the four deoxyribonucleoside 5'-triphosphates, one of which (X) is labelled with ^{32}P. The new polynucleotide is degraded using an enzyme which severs the 5'-phosphate linkages. The ^{32}P is then attached to the nucleoside adjacent to the position of incorporation of X. Separation and study of distribution of ^{32}P amongst the nucleoside 3'-phosphates enables the frequencies of all possible 16 dinucleotide pairs to be determined, after suitable checks and corrections have been made. The frequencies of adjacent purine nucleotides were non-random and agreed with those determined chemically, but otherwise the results cannot be directly compared, since different sequences are examined. The results from this method, as already mentioned, also confirm the opposite directions of the two chains in DNA and the AT and GC pairing schemes.

It is useful to compare the information on sequences in DNA from organisms selected because of their biological relationship, or lack of it. Thus DNA from two related bacterial species (*E. coli* and *Salmonella typhimurium*) have closely similar frequencies of occurrence of pyrimidines, whereas DNA from two unrelated species (*Pseudomonas aeruginosa* and *Alcaligenes faecalis*) differ greatly in this respect[41].

A quite different approach to this problem has now become available with the discovery that, under carefully chosen conditions, the two strands of DNA of micro-organisms and bacteriophage may be separated and then subsequently reunited to restore the original helical structure and biological activity[42]. This process of 'renaturation' may be followed by centrifugation of DNA in a density gradient of caesium chloride[43]. The DNA is concentrated into a band at a certain position in the centrifuge cell dependent upon its density. By appropriate labelling of the DNA with ^{15}N or deuterium, strands from a given DNA may be characterized by their density, as may new double helices formed by the union of two strands from originally different DNA, one labelled and the other unlabelled. (This technique is discussed further in Chapter 8 in connection with the mechanism of replication of DNA.)

The ability of two polynucleotide strands to unite to form a DNA double helix is dependent on the complementarity of the nucleotide sequences according to the AT and GC matching rule. Thus the ability of two strands to unite when these are derived from DNAs of different organisms is a measure of the similarity of the nucleotide sequences in the two original DNAs. The similarity between the base composition of organisms related genetically and taxonomically has been shown to extend also to their ability to form DNA hybrids, that is, to the similarity of their DNA nucleotide sequences[44]. Thus the organism *B. subtilis*, as representative of the Bacillaceae, was shown to be both genetically related (with respect to its ability to be transformed by DNA from other Bacillaceae) to certain other organisms of this group and to be capable of forming DNA hybrid bands with the same organisms[44]. Such a close correlation clearly has great potentialities in the assessment of the relationship between micro-organisms but cannot be used for higher plants and animals since their DNA are more heterogeneous and do not 'renature'. The method, while showing that the nucleotide sequences of certain DNA are at least partially homologous, does not, of course, yield any direct information about the actual sequences themselves. It may in some instances be more sensitive than chemical methods since, for example, the DNAs from *E. coli* and *Sal. typhimurium* do not readily form a molecular hybrid[42], although the chemical evidence suggests a similarity of sequence[41]. Similarity of GC content is not by itself a guarantee of the ability to form hybrids: thus *B. subtilis* and *B. brevis* have the same GC content but do not form hybrids and, indeed, are not genetically related, although they belong to the same taxonomic group[44].

ENZYMATIC SYNTHESIS

The complementary structure of DNA implies that in a system in which DNA is being formed the new DNA molecules are in some way built up on the older ones. This implication has received striking confirmation from the work of Kornberg and his colleagues, who have been able to purify from an extract of *E. coli* an enzyme ('polymerase') which catalyses the incorporation of deoxyribonucleotides into DNA[45]. The nucleotides had to be in the form of the nucleoside triphosphates and it was found that for the most rapid incorporation into DNA the triphosphates of all four of the deoxyribonucleosides which commonly occur in DNA (A, T, G, C) had to be present in the reaction mixture at 37° C, together with magnesium ions.

The synthetic reaction is summarized by Kornberg[45] as

$$n \begin{bmatrix} \overset{*}{\text{TPPP}} \\ \overset{*}{\text{dGPPP}} \\ \overset{*}{\text{dAPPP}} \\ \overset{*}{\text{dCPPP}} \end{bmatrix} + \text{DNA} \rightleftharpoons \text{DNA}\begin{bmatrix} \overset{*}{\text{TP}} \\ \overset{*}{\text{dGP}} \\ \overset{*}{\text{dAP}} \\ \overset{*}{\text{dCP}} \end{bmatrix}_n + 4(n)\text{PP},$$

where the symbols in the left-hand bracket represent the triphosphates of the deoxyribonucleosides and those in the right-hand bracket the deoxyribonucleotides; PP = pyrophosphate and $\overset{*}{P}$ = phosphoric residue incorporated into the DNA, which in some experiments was labelled with ^{32}P. Increases of DNA by a factor as high as 10–20 were

38

obtained when all four triphosphates were present, so that 90–95 per cent of the DNA isolated from the reaction mixture was derived from them. It was found, significantly, that the presence of polymerized DNA was essential for this net synthesis of new DNA, which recalls the hypothesis of DNA being a necessary 'template' for the production of new DNA. The analogy of a template was further justified by the observations that the newly synthesized DNA had a composition characteristic of the double-helical DNA structure (namely $A = T$, $G = C$) and that the ratio $(A+T)/(G+C)$ in the new DNA corresponded very closely to the ratio present in the DNA primer used to initiate its synthesis, whatever the relative composition of the reaction mixture. This was tested with DNA primers of an unusually wide range of composition with $(A+T)/(G+C)$ varying from $0.5 > 40$, since it was also possible to use a copolymer of deoxyadenylate and thymidylate $((A+T)/(G+C) > 40)$, which was obtained from the polymerase system when DNA was absent. Once formed and isolated this AT polymer initiated the synthesis of new AT polymer, despite the presence in the polymerase reaction mixture of the nucleoside triphosphates of cytosine and guanine, in addition to those of adenine and thymine.

The concept of DNA as a template is further supported by the observation that the molecular weight of the DNA synthesized was about 5×10^6 and approximately equal to that of the calf thymus DNA used as primer. Moreover, it had the characteristics of the double-helical configuration and its polydeoxyribonucleotide chains were joined by $3', 5'$ diester bonds. Thus it seems clear that the new DNA is formed on the pattern of the old.

It is known that various analogues of the purine and pyrimidine bases may be incorporated into bacterial and viral DNA so it is not unexpected that similar substitutions can be made in enzymically synthesized DNA. Thus, in the synthetic system, uracil and 5-bromouracil can be incorporated into DNA instead of thymine, 5-methyl and 5-bromocytosine instead of cytosine, and hypoxanthine (but not xanthine) instead of guanine. The 'unnatural' bases which can be substituted all have groups at the 1 and 6 positions capable of forming bonds at the positions postulated by Watson and Crick.

Even when only one deoxyribonucleoside triphosphate is present this can be incorporated into the DNA primer and it becomes attached to the free nucleotide end of the DNA molecule by a $3', 5'$ diester linkage. It is not clear how the mechanism of this limited reaction which lengthens the DNA strand is related to the extensive synthetic reaction in which the *average* DNA chain length remains unchanged

39

and in which replicated double-helical molecules are formed. The results seem difficult to accommodate to the 'Y' scheme of Watson and Crick since a new nucleotide will only add on to that strand which terminates in a nucleotide[46]. If it is only singly stranded regions of the DNA which act as the priming site, as seems possible, this would provide a common basis for both the limited and the synthetic reactions. Thus these studies on DNA synthesis have had the very important result of providing evidence that DNA acts as a template at the molecular level, since they have shown that new DNA is synthesized according to a molecular pattern determined by that of the DNA already present.

REFERENCES

[1] Levene, P. A. and Bass, L. W. *Nucleic Acids.* New York; Chemical Catalog Co., 1931

[2] Brown, D. M. and Todd, A. R. In *The Nucleic Acids* (Ed. Chargaff and Davidson), Vol. 1, p. 409. New York; Academic Press, 1955

[3] Jordan, D. O. *The Chemistry of Nucleic Acids.* London; Butterworths, 1960

[4] Dunn, D. B. and Smith, J. D., *Nature, Lond.*, 1954, **174**, 305; 1955, **175**, 336

[5] Peacocke, A. R. *Chem. Soc. Special Publ.* 1957, No. 8, p. 139

[6] Chargaff, E. In *The Nucleic Acids* (Ed. Chargaff and Davidson), Vol. 1, p. 307. New York; Academic Press, 1955

[7] Sueoka, N. *J. mol. Biol.*, 1961, **3**, 31

[8] Lee, K. Y., Wahl, R. and Barbu, E., *Ann. Inst. Pasteur*, 1956, **91**, 212

[9] Belozersky, A. N. and Spirin, A. S., *Nature, Lond.*, 1958, **182**, 111

[10] Gulland, J. M., Jordan, D. O. and Taylor, H. F. W. *J. chem. Soc.*, 1947, 1131

[11] Chargaff, E., Crampton, C.F. and Lipshitz, R. *Nature, Lond.*, 1953, **172**, 289

[12] Hamilton, L. D., Barclay, R. K., Wilkins, M. H. F., Brown, G. L., Wilson, H. R., Marvin, D. A., Ephrussi-Taylor, H. and Simmons, N. S. *J. biophys. biochem. Cytol.*, 1959, **5**, 397

[13] Sinsheimer, R. L. *J. biol. Chem.*, 1955, **215**, 579

[14] Marmur, J. and Doty, P. *Nature, Lond.*, 1959, **183**, 1427; Sueoka, N., Marmur, J. and Doty, P. *Nature, Lond.*, 1959, **183**, 1429; Rolfe, R. and Meselson, M. *Proc. nat. Acad. Sci., Wash.*, 1959, **45**, 1039

[15] Wilkins, M. H. F., Stokes, A. R. and Wilson, H. R. *Nature, Lond.*, 1953, **171**, 738; Franklin, R. E. and Gosling, R. G. *Nature, Lond.*, 1953, **171**, 740

[16] Watson, J. D. and Crick, F. H. C. *Nature, Lond.*, 1953, **171**, 737; Watson, J. D. and Crick, F. H. C. *Proc. Roy. Soc. A*, 1954, **223**, 80

[17] Wilkins, M. H. F., *J. Chim. phys.*, 1961, **58**, 891; Marvin, D. A., Spencer, M., Wilkins, M. H. F. and Hamilton, L. D. *Nature, Lond.*, 1958, **182**, 387

[18] Josse, J., Kaiser, A. D. and Kornberg, A. *J. biol. Chem.*, 1961, **236**, 864; Swartz, M. N., Trautner, T. A. and Kornberg, A. *J. biol. Chem.*, 1962, **237**, 1961

[19] Feughelman, M., Langridge, R., Seeds, W. E., Stokes, A. R., Wilson, H. R., Hooper, C. W., Wilkins, M. H. F., Barclay, R. K. and Hamilton, L. D. *Nature, Lond.*, 1955, **175**, 834

REFERENCES

[20] Shooter, K. V. *Progr. Biophys.*, 1957, **8**, 310
[21] Peacocke, A. R. *Progr. Biophys.*, 1960, **10**, 55
[22] Luzzati, V. *J. Chim. phys.*, 1961, **58**, 899
[23] Doty, P. *Rev. mod. Phys.*, 1959, **31**, 107
[24] Donohue, J. *Proc. nat. Acad. Sci., Wash.*, 1956, **42**, 60
[25] Miles, H. T. *Proc. nat. Acad. Sci., Wash.*, 1961, **47**, 791
[26] Pauling, L. and Corey, R. B. *Arch. Biochem.*, 1956, **65**, 164
[27] Cox, R. A., Wilson, S., Overend, W. G. and Peacocke, A. R. *Nature, Lond.*, 1955, **176**, 919
[28] Pullman, B. and Pullman, A. *Nature, Lond.*, 1961, **189**, 725
[29] Herskovits, T. T., Singer, S. J. and Geiduschek, E. P. *Arch. Biochem.*, 1961, **94**, 99
[30] Hamilton, L. D., Barclay, R. K., Wilkins, M. H. F., Brown, G. L., Wilson, H. R., Marvin, D. A., Ephrussi-Taylor, H. and Simmons, N. S. *J. biophys. biochem. Cytol.*, 1959, **5**, 397
[31] Cavalieri, L. F. and Rosenberg, B. *Biophys. J.*, 1961, **1**, 317, 323, 337
[32] Cairns, J. *Nature, Lond.*, 1962, **194**, 1274
[33] Sinsheimer, R. L., *J. mol. Biol.*, 1959, **1**, 43: Fiers, W. and Sinsheimer, R., *J. mol. Biol.*, 1962, **5**, 408, 420, 424
[34] Levinthal, C. and Thomas, C. A. In *The Chemical Basis of Heredity* (Ed. McElroy and Glass), p. 737. Baltimore; Johns Hopkins Press, 1957
[35] Davison, P. F. *Proc. nat. Acad. Sci., Wash.*, 1959, **45**, 1560; *Nature, Lond.*, 1960, **185**, 918
[36] Rubenstein, I., Thomas, C. A. and Hershey, A. D. *Proc. nat. Acad. Sci., Wash.*, 1961, **47**, 1113; Davison, P. F., Freifelder, D., Hede, R. and Levinthal, C., *Proc. nat. Acad. Sci., Wash.*, **47**, 1123; Cairns, J. *Mol. Biol.*, 1961, **3**, 756
[37] Hershey, A. D., Burgi, E., Cairns, H. J., Frankel, F. and Ingraham, L. *Yearbook 60, Carnegie Institution of Washington*, p. 455. 1960–61
[38] Cairns, J. *J. mol. Biol.*, 1962, **4**, 407
[39] Kleinschmidt, A., Lang, D. and Zahn, R. K. *Z. Naturf.*, 1961, **16b**, 730
[40] Shapiro, H. S. and Chargaff, E. *Biochim. biophys. Acta*, 1957, **23**, 451; Burton, K. *Biochem. J.*, 1960, **74**, 35P; Burton, K. and Petersen, G. B. *Biochem. J.*, 1960, **75**, 17; Jones, A. S. and Stacey, M. *Chem. Soc. Special Publ.*, No. 8, p. 129; Sinsheimer, R. L. *J. biol. Chem.*, 1955, **215**, 579
[41] Burton, K. *Brit. med. Bull.*, 1962, **18**, No. 1, 3
[42] Doty, P., Marmur, J., Eigner, J. and Schildkraut, C. *Proc. nat. Acad. Sci., Wash.*, 1960, **46**, 461
[43] Meselson, M., Stahl, F. W. and Vinograd, J. *Proc. nat. Acad. Sci., Wash.*, 1957, **43**, 581
[44] Marmur, J., Schildkraut, C. L. and Doty, P. *J. Chim. phys.*, 1961, **58**, 945; Marmur, J., Rownd, R. and Schildkraut, C. L. *Progress in Nucleic Acid Research* (Ed. Davidson and Cohn), Vol. 1, New York; Academic Press
[45] Bessman, M. J., Lehman, I. R., Adler, J., Zimmerman, S. B., Simms, E. S. and Kornberg, A. *Proc. nat. Acad. Sci., Wash.*, 1958, **44**, 633; Adler, J., Lehman, I. R., Bessman, M. J., Simms, E. S., and Kornberg, A. *Proc. nat. Acad. Sci., Wash.*, 1958, **44**, 641; Lehman, I. R., Zimmerman, S. B., Adler, J., Bessman, M. J., Simms, E. S. and Kornberg, A. *Proc. nat. Acad. Sci., Wash.*, 1958, **44**, 1191
[46] Delbruck, M. and Stent, G. S. *The Chemical Basis of Heredity* (Ed. McElroy and Glass), p. 694. Baltimore; Johns Hopkins Press, 1957

41

CHAPTER 5

RIBONUCLEIC ACID

LOCATION AND CLASSIFICATION

RNA occurs in the cytoplasm of all cells and is present in particularly large amounts in cells in which rapid protein synthesis is occurring, either for secretion or growth (Chapter 11). Its presence has been demonstrated[1] by chemical methods, by ultra-violet spectroscopy and by staining methods, combined with the action of ribonuclease to distinguish between DNA and RNA. Relatively small amounts of RNA have also been detected in the nucleolus of the nucleus.

As will be discussed in more detail when the mechanism of protein synthesis is examined (Chapter 11), RNA may be classified into three main types.

(1) Soluble or amino acid transfer RNA can be linked to an amino acid by a covalent acyl bond between the amino acid carboxyl group and one of the ribose hydroxyl groups of the adenylic acid residue which terminates the RNA polynucleotide chain. Each transfer RNA molecule contains about 80 nucleotides and conveys a specific amino acid to the ribosomes where it is then utilized in protein synthesis.

(2) Macromolecular RNA (molecular weight about 10^5–10^7) includes both the RNA of ribosomes and that of viruses, since these two types of RNA have much the same properties in free solution and for a long time it was thought to be the only type of RNA. Viral and ribosomal RNA differ in their mode of participation in their parent structures and in their function, since viral RNA is a genetic carrier whereas ribosomal RNA is not.

(3) Messenger RNA is a short-lived intermediate with a rapid turnover which carries genetic information from the DNA gene to the ribosomes in which protein is actually synthesized.

These three types of RNA all have the same covalent chemical structure, but their macromolecular properties are sufficiently distinct to merit separate treatment. Only such structural questions will be considered in this chapter; their biological function is described later.

CHEMICAL STRUCTURE

As with DNA, the component units of RNA have been deduced from extensive studies of chemical hydrolysis and enzymic degradation[2-5]. Acid hydrolysis yields adenine, guanine, D-ribose (XI), inorganic phosphate and pyrimidine nucleotides which slowly break down to inorganic phosphate and pyrimidine nucleosides. Under more drastic conditions these latter yield the bases cytosine and uracil (XII). The

XI   XII

purine nucleotides are more stable to alkali, and alkaline hydrolysis of RNA yields a mixture of the mononucleotides of the four bases. Under appropriate conditions, purine and pyrimidine nucleosides together with inorganic phosphate can be obtained from these nucleotides. RNA can be broken down by the enzyme ribonuclease to give pyrimidine, but not purine, mononucleotides and a mixture of oligonucleotides which are relatively richer in purines than the original RNA. The detailed mechanism of the enzymic process has been much examined[4], but for the present purpose it is sufficient to state that from these degradation studies the various components of the RNA can be related in the following way:

The bases which occur in RNA are mainly the purines, adenine and guanine, and the pyrimidines, cytosine and uracil. This list differs from that for DNA only in that uracil replaces thymine. It has been known for some time[6] that small amounts of thymine, methylated adenines, methylated guanines and pseudo-uracil (5-ribosyl uracil: ribose attached to the uracil ring at position 5 by a C—C bond) occur as part of the structure of some natural RNAs, though not of viral RNA. These rarer bases never occur to the extent of more than 4 per cent, and usually less than this. As described in the next section, soluble RNA contains a number of unusual bases, some of which have only recently been fully identified.

RNA differs from DNA, of course, in containing the sugar D-ribose (XI) in its furanose form instead of 2-deoxy-D-ribose (V, Chapter 4). These sugars differ at position $2'$ of the ring, where the ribose has a hydroxyl group and the deoxyribose has a hydrogen atom; this causes a marked difference in their chemical behaviour[7]. It prevents the formation in DNA of a cyclic diester between the $2'$ and $3'$ positions on the sugar ring under the influence of alkali and thereby stabilizes the DNA to this reagent relative to RNA, which is readily hydrolysed by alkali. The sites of linkage of the bases to the sugar in RNA are the same as in DNA, that is, at N_9 of the purines and N_3 of the pyrimidines.

As with DNA, the chemical investigation of the process of degradation and of the nature of the various breakdown products[5], together with the titration evidence[8] have led to the formulation of a recurring $3',5'$-internucleotidic linkage for the main chain of RNA (as in VII, Chapter 4; apart from the extra hydroxyl at $C_{2'}$ of the sugar, formula VI, Chapter 4 can also represent RNA). In spite of earlier indications to the contrary, it is now clear that undegraded RNA is unbranched, although there are two places ($2'$ and $3'$) at which each ribose ring might be esterified. The evidence for this is partly chemical[4] and partly based on a more refined analysis[8] of the titration curves of RNA, which shows no significant proportion of the secondary phosphoryl groups which should occur at the more numerous chain ends of a branched structure.

SOLUBLE RNA (AMINO ACID TRANSFER RNA*)

The existence of a soluble RNA (S-RNA*) fraction with the specific property of binding amino acids is now well established[9]. The formation of each compound between an amino acid and RNA is catalysed

* Frequently now denoted as t-RNA, especially when the RNA which transfers a particular amino acid is in view.

by a specific enzyme. In these compounds the carboxyl group of the amino acid is linked to the ribose hydroxyl group on the adenylic acid residue at the end of the chain of the soluble RNA, so-called because it represents a fraction of RNA not readily precipitated and easily extracted by various reagents. Molecular weights of 24,000–26,000 have been obtained by sedimentation, diffusion and viscosity measurements[10] and indicate very little heterogeneity. Nucleotide analyses[11] show that the composition is quite different from that of ribosomal RNA from the same cells and usually imply a minimum molecular weight of about 20,000 to 30,000, about 60 to 100 nucleotides.

Individual S-RNAs, specific for the transfer of a particular amino acid have been isolated and purified[12, 13], for example, alanine, tyrosine and valine transfer RNAs, and two different RNA molecules (I and II) which transfer serine. Degradation of these S-RNA and separation of their constituent nucleotides has revealed that they contain a number of unusual nucleotides[12, 14]. In alanine transfer RNA, there occurs, in addition to the usual 3′-nucleotides of adenine (A), guanine (G), cytosine (C) and uracil (U), the following nucleotides: 4,5 dihydrouridine 3′-phosphate (DiHU); N(2)-dimethylguanosine 3′-phosphate (DiMeG); inosine 3′-phosphate (I; inosine has the same structure as the ribonucleoside, guanosine, but without the 2-amino group on the guanine); 1-methyl inosine 3′-phosphate (MeI); 1-methyl guanosine 3′-phosphate (MeG); pseudo-uridine 3′-phosphate (ψ, the ribonucleotide of 5-ribofuranosyluracil, see XII and page 44); and ribothymidine 3′-phosphate(rT), the ribonucleotide of thymine, with ribose instead of the deoxyribose of thymidine 3′-phosphate. The serine transfer RNAs (both I and II) contain all the above, except MeG and MeI, and the following additional nucleosides, as nucleotides, with the phosphate at the 3′ position: N(6)-(γ,γ-dimethylallyl) adenosine or isopentenyladenosine (iPA); 5-methyl cytidine (MeC); N(6)-acetyl cytidine (AcC); 2′-O-methyl guanosine (OMeG); and 2′-O-methyl uridine (OMeU).

Tyrosine transfer RNA contains, besides A, U, G and C, the nucleotides DiHU, DiMeG, ψ, rT, MeC, OMeG, and a monomethylated (MeA) and dimethylated adenine (DiMeA). Within a margin of a few nucleotides, analysis of the total S-RNA of yeast, E. coli and rat liver showed that $G \simeq C$ and $A \simeq U$, which corresponds to the relation for double-helical DNA, since uracil occurs in RNA instead of thymine in DNA. Since uracil can be involved in the same type of hydrogen bonding as thymine (IX, X, page 32), it was naturally inferred that much of S-RNA was double-helical. End-group analysis and molecular weight studies showed that S-RNA possessed a single polynucleotide chain[15]. The changes of ultra-violet absorption with temperature and other studies[10, 16] were interpreted to mean that there is a single, long double-helical region exhibiting DNA-type base-pairing between complementary base sequences. However, this later proved to be an over-simplification of the structure which could only be precisely defined when the actual base sequence had been determined. (An x-ray study[17] of a crystalline, but degraded, RNA which largely consisted of double helices and which was at first thought to be S-RNA, later transpired[18] to be derived from ribosomal RNA and will be discussed below in that context.)

The first sequence studies revealed[19] that, at the 3'-phosphate end which accepts the amino acid, the sequence is always CCA, that is cytidylic, cytidylic, adenosine 3'-phosphate. At the 5'-phosphate end, a G residue was found[20] and has been confirmed in alanine[21], serine[22,23] and valine[24,25] transfer RNAs but not in tyrosine[26] transfer RNA which has a cytidylic acid residue at this end.

That each S-RNA chain specific for the transfer of a particular amino acid has a unique base sequence became a possibility when it was shown that the sequences of bases adjacent to the terminal CCA group were different in leucine and iso-leucine transfer RNAs[27] and that the arrangement of bases was non-random throughout the whole sequence of the average S-RNA molecule[15a]. This idea was convincingly ratified when Holley et al.[21] brought to fruition many years of careful work on the enzymatic degradation of RNA by proving for the first time the base sequence of a natural nucleic acid, that of alanine transfer RNA from yeast.

Structure of alanine transfer RNA

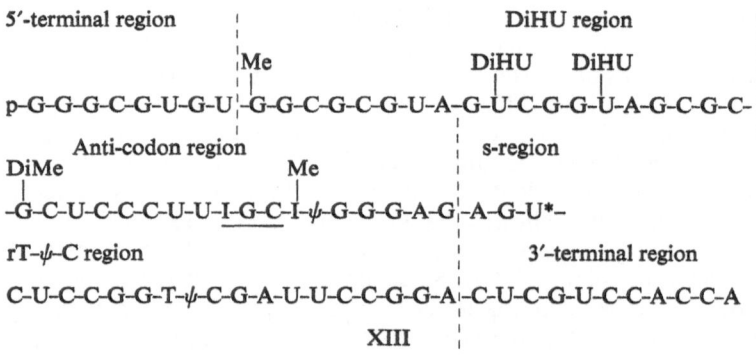

XIII

The structure (XIII) is usually written with the free 5'-hydroxyl of the ribose on the left, and the free 3'-hydroxyl on the right. This achievement involved the identification[14] of small fragments formed by complete digestion of the RNA with pancreatic ribonuclease, which cleaves the RNA next to pyrimidine nucleotides (e.g. C– and U–); and with takadiastase ribonuclease T1, which cleaves the RNA chain next to G– and I–. Assembly of these structures yielded a description of the sequence in terms of 16 oligonucleotides, containing altogether 77 nucleotide residues, of a total molecular weight of 26,600, as the sodium salt. Degradation of the alanine transfer RNA by takadiastase ribonuclease T1 under controlled conditions (e.g. 1 hour at 0°C) yielded a number of larger oligonucleotide fragments, the analysis of which afforded the correct·order of assembly of the smaller oligonucleotides in the RNA and thence a unique base sequence for it. Certain features of this structure were clear at once. There was no obvious pattern in the distribution of the minor or unusual nucleotide residues. The alanine code-word or 'codon' (see page 178 and Glossary) in messenger RNA is now known to be fourfold degenerate (that is, GCX, where X = U, C, A or G) and the code is read in the 5' to 3' direction. The coding triplet or 'anti-codon' on the S-RNA which is complementary to this codon is CGY in the 3' to 5' direction, i.e., YGC in the 5' to 3' direction of the structure XIII. Here Y represents a third nucleotide of the anti-codon which initially was presumed to be able to complement X, but since X can be any of the four usual bases the pairing at this point seems to be less critical than at the other two positions. (N.B. Inosine (I) appears to be equivalent to G in complementary pairing of the Watson-Crick type (IX, X, page 32), i.e., I pairs with C; and U is equivalent to T, i.e., U pairs with A). The sequence YGC occurs at 6 places in the sequence XIII but the choice between these for the anti-codon is limited by (*i*) whether or not the particular sequence YGC is already involved in base pairing of the Watson-Crick type in short double-helical runs, and by (*ii*) considerations about which pairs XY are structurally possible.

When the alanine S-RNA structure XIII is examined to find which structures are permitted with G–C and A–U complementary base-pairs, it is found that the longest complementary sequence contains only 5 base pairs, contrary to previous ideas[15b, 17]. *Figure 5.1* shows one of the most likely of these structures (the 'clover-leaf'). This has, at bases numbered 36 to 38, the sequence I–G–C, which is located in a single-stranded loop and, since it is not involved in intramolecular base-pairing, would be available as a suitable anti-codon for the alanine codon on the messenger RNA, as first suggested by Holley *et al.*[21]. This suggestion has been greatly strengthened by the subsequent determination of the sequences of other transfer RNAs and the realization that these can also take up a clover-leaf structure, with bases numbered 36 to 38 present in a loop and also being suitable anti-codons for the amino acid for which the transfer RNA is specific. These relationships are summarized in Table 5.1 and the 'anti-codon

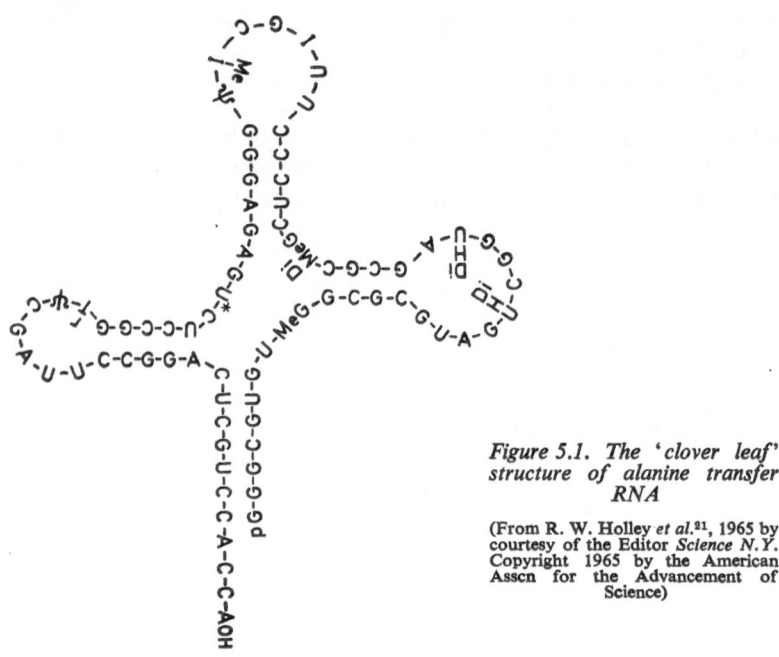

Figure 5.1. The 'clover leaf' structure of alanine transfer RNA

(From R. W. Holley *et al.*[21], 1965 by courtesy of the Editor *Science N.Y.* Copyright 1965 by the American Assen for the Advancement of Science)

region' of the transfer RNAs in such clover leaf structures as XIII are illustrated in *Figure 5.2*.

So far only the bearing of the secondary structure of transfer RNAs on the location of the anti-codon has been discussed. But this is also determined by the nature of the third nucleotide, for example, Y in the sequence YGC, the anti-codon for alanine. Crick[29] has proposed that, at this third position, the structural requirements are less strict than those of the usual complementary base pairs (A–U, G–C) so that U on the anticodon recognizes A or G, C recognizes G, A recognizes U,

G recognizes U or C and I recognizes U, C, or A, on account of the slight 'wobble' allowed in the matching structures at this position. There is some direct experimental confirmation[30] of these suggestions, which are in accord with the relationship summarized in Table 5.1.

The evidence from the S-RNA sequences so far determined suggests other similarities between the sequences in alanine, serine and tyrosine

Anticodon region

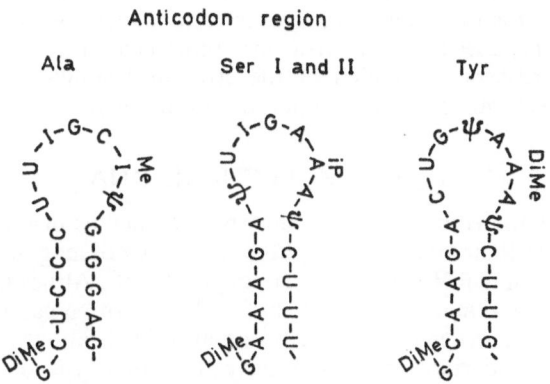

Figure 5.2. The anti-codon region in the clover-leaf structure for alanine, serine (I and II) and tyrosine transfer RNAs[23]. The anti-codon triplet is thought to consist of the top 3 nucleotides in the loop (compare Table 5.1)

(From H. Zachau[23], 1966 by courtesy of the Editors of *Cold Spring Harb. Symp. on Quantitative Biology*)

TABLE 5.1

Transfer RNA specific to	Codon(s)	Anti-codon type	Possible actual anti-codon sequence	Position	Ref.
Alanine	GCX	–GC	IGC	36–38	21
Serine (I and II)	UCX	–GA	IGA	36–38	22, 23
	AGZ	–CU	(in both I and II)		
Tyrosine	UAZ	–UA	*GψA	36–38	26
Valine	GUX	–AC	IAC	38–40(?)	25, 28

Codon and anti-codon sequences read from 5′ on left to 3′ on right.
X = U, C, A or G
Z = U or C
* ψ, like U, is complementary to A.

4   49

transfer RNAs. These include[23]: the similar distances between two of the DiHUs; the common sequence –G–C–DiMeG– and the anticodon region; the anti-codon in all three is flanked by –U– on the 5'-side; in the –rT–ψ–C– region the distance between Cp or MeCp at the 5'-side and the rTp is identical (–p means 3'-phosphate); serine and tyrosine transfer RNAs are very similar in the DiHU and anticodon regions. The common ability of these RNAs to form the clover leaf pattern is suggestive since, apart from the specific region for recognizing the codon (the anti-codon) in the messenger RNA, each S-RNA must also have a region which recognizes the aminoacyl S-RNA synthetase (Chapter 11) specific for a particular amino acid and a region which binds to ribosomes. Which of the structural similarities and dissimilarities is related to these functions is unknown.

MACROMOLECULAR RNA

Both viral and ribosomal RNA may be classified as macromolecular RNA, since their molecular weights are an order of magnitude greater than those of S-RNA, being in the range $10^5–10^7$. Although differing in function and in molecular weight (viral RNA often has a molecular weight of the order 10^7, while ribosomal RNA is about $5 \times 10^5–10^6$) both viral and ribosomal RNA have the same type of configuration in solution and they will, therefore, be considered together. However, the relationship between the configuration of these two kinds of macromolecular RNA in solution to that in the original biological structure (virus or ribosome) is not the same and must be considered later.

Chemical analysis of RNA has not led to the clear generalizations concerning base ratios that have been found with DNA. For a long while it proved difficult to isolate RNA samples of reproducible composition from sources other than viruses. This difficulty arose not only from the stability and activity of ribonuclease during extraction, but also from the presence, as we now know, of at least three different types of RNA of distinct functions in protein synthesis. Moreover, it was observed in many laboratories that the last traces of protein were held very tenaciously as a complex with the RNA. The ratio of purines to pyrimidines varies considerably and does not often equal unity, the value found with DNA[31]. The analyses could be criticized as not referring to the total RNA in the cell, so techniques were developed for analysing the total RNA in viruses or cells, while the RNA was still attached to protein[31,32]. Under these conditions only one regularity has been observed, namely that the number of 6-amino groups is equal to the number of 6-oxo groups, the groups which in DNA are involved in hydrogen bonding between base pairs. Thus, $A + C = G + U$. (This does not apply to plant virus RNA.) Thus it seems unlikely that there can be complete complementary pairing (A to U, G to C) between two chains in RNA, although partial pairing is not excluded. Elson and Chargaff[32] suggested that the equality of 6-amino and 6-oxo groups in RNA when it is not separated from protein might be attributed to hydrogen bonding to the protein. The 6-amino group of adenine or cytosine in one polyribonucleotide chain might be hydrogen bonded to the $\rangle C{=}O$ group of the peptide bond of the protein and a 6-oxo group in another chain (or another part of the same chain) likewise hydrogen bonded to the —NH— of the peptide linkage. This would ensure the equality observed. However, there is no other evidence for this proposal and the known structures of ribonucleoproteins do not support it.

Species-specificity of base composition has been observed in the RNA of plant viruses, while different strains of the same species have an identical composition[33]. The composition of RNA in microorganisms[34] varies much less than does the DNA from the same organisms: thus, for RNA, $(A+U)/(G+C)$ is $1\cdot03$–$1\cdot45$, but for DNA, $(A+T)/(G+C)$ is $0\cdot37$–$2\cdot2$. This constitutes a problem for understanding the genetic code as will be discussed later (Chapter 12). Yeast, animal and viral RNA differ in composition from one another[31] but it is now clear that such generalizations have little significance unless the function of the RNA is more closely defined. Separation and identification of the mixture of nucleotides produced by ribonuclease action on RNA reveal that pyrimidine nucleotides tend to be 'bunched' together[35], so that there is a more than average chance that a pyrimidine will be located next to another pyrimidine. It follows that a similar relation must also apply to adjacent purine nucleotides. By such methods it has been shown that the extent of bunching of pyrimidines differs in the RNA of three strains of TMV, although the RNAs have an identical total composition[36]. Hence, the sequence of the nucleotides in these three strains must be different. This observation constituted one of the earliest direct indications of the biological significance of the sequence of bases in the nucleic acids.

Until the last few years, different preparations of RNA, even from the same source, were very heterogeneous with respect to molecular weight[37]. Some of this heterogeneity can now be attributed to the presence in the original cells of RNA of different sizes and functions, but much more can be attributed to degradation by ribonuclease or by the alkali or heat employed in some isolation procedures. Only from about 1951 onwards have molecular weights above 10^5 been reported for RNA. The RNA which has been the best characterized in recent years is that of TMV through refinement of preparative methods which have been checked by its molecular weight and by the biological activity of the RNA, once its independent infectivity had been established (*see* Chapter 3). Homogeneous RNA of molecular weight of about $2 \cdot 0 \times 10^6$ have thus been obtained[38] and this is not far from the value of $2 \cdot 5 \times 10^6$, which is all the RNA in an individual TMV particle.

There has also been an improvement in the preparation of ribosomal RNA. This has depended firstly on a recognition of the conditions governing the stability of the various microsomal particles which may be obtained. These particles are usually characterized by their approximate sedimentation coefficients. Particles of sedimentation coefficient $30S$ and $50S$ can combine under the influence of magnesium ions to form a $70S$ particle, two of which may aggregate to give a $100S$ particle[39]. It appears [38,39] that from the $30S$, $50S$ and $70S$ particles may be obtained RNA of sedimentation coefficient about $15-18S$ and molecular weight of about 5×10^5. From the $50S$ and $70S$ microsomal particles is also obtained RNA of sedimentation coefficient about $23-32S$ and molecular weight of about 10^6. These observations[40-43] have been made on RNA derived mainly by the phenol method of extraction[44] from a wide variety of sources, bacterial, plant and animal, including ascites tumour cells. The investigations seemed to indicate that one molecule of an RNA of molecular weight $5-6 \times 10^5$ associated with another RNA of molecular weight $1 \cdot 0-1 \cdot 3 \times 10^6$ to form a single microsomal particle, but the matter was complicated by the apparent tendency of ribosomal RNA to be degraded into smaller fragments when exposed to elevated temperature, dialysis and certain proteins. However, it was eventually shown[44,45] that all of these effects are produced by traces of ribonuclease and that there is no 'sub-unit' RNA of a limiting molecular weight as low as 120,000 into which ribosomal RNA was once thought to dissociate. It has been claimed[46] that it is the RNA of molecular weight about 5×10^5 ($16S$) which is the true sub-unit and that this can associate by lateral aggregation into dimers of molecular weight about 10^6 ($23S$), and even trimers ($30S$) but this view would not now command general support.

The problem of the configuration of macromolecular RNA only became soluble as preparations improved. The earlier[47] x-ray diffraction patterns of oriented RNA fibres were diffuse and had not led to any clear-cut result; but they had shown regular spacings, like

DNA, at 3·3, 4·0, and 25 Å along the axis, which suggested the possibility that RNA might contain a double-helical structure of the DNA type.

However, the undegraded RNA which was isolated from TMV appeared to consist of only one polynucleotide strand. Thus, inactivation of the virus by heat was a consequence of the cleavage of only one internucleotide bond in a chain[48] and a similar conclusion could be drawn from a study of ribonuclease action[49]. This behaviour contrasts with that of DNA which requires two breaks, one in each of the two strands, to cause a change in molecular weight (Chapter 4) and a loss of biological activity[50]. Moreover, x-ray diffraction studies have shown that RNA is present in the intact TMV as a single helical chain within a protein matrix (Chapter 6), so that there is a strong presumption that the chain remains single when free in solution.

The behaviour of macromolecular RNA in solution, whether viral or ribosomal, is that of a flexible charged macromolecule, which uncoils reversibly as the ionic strength is lowered. At relatively low ionic strength it obeys the relationships between molecular weight and sedimentation coefficient and viscosity characteristic of a randomly coiled single chain structure with many intramolecular contacts[37, 51]. The actual extent of coiling of macromolecular RNA in solution is greater and its rate of change with decreasing ionic strength is somewhat less than that expected for a charged macromolecule, so that intramolecular attractions within the single chain must play a considerable part in determining its configuration.

The earliest evidence for a secondary structure in RNA came, as with DNA, from titration studies. Macromolecular RNA from *Aerobactor aerogenes* and from yeast exhibited a small, but definite, hysteresis in its titration curves to and from the acid and alkaline regions of pH[52], a phenomenon which, in DNA, is associated with the presence of Watson–Crick hydrogen bonding. A titration hysteresis has also been observed[53] with RNA from *E. coli*, when it was shown, unlike DNA, to be a *reversible* phenomenon which disappeared at 38° C. A reversible transition between a hydrogen-bonded and non-hydrogen-bonded state was thus indicated.

The disappearance of the hysteresis with rise in temperature is consistent with many other observations on the changes in ultra-violet absorption and in optical rotation which occur when solutions of macromolecular RNA are heated[42, 54, 55]. These changes occur over a wide temperature range and are more reversible than with DNA, which shows that the ordered, hydrogen-bonded sections of the chain are shorter and more defective than in DNA. Detailed examination of the changes and comparison with similar changes in synthetic poly-ribonucleotides show that a large proportion (one estimate[55] is about 80 per cent) of the bases in viral and ribosomal RNA are hydrogen bonded in pairs (A to U, G to C) in short double-helical regions formed between different parts of the same single strand of RNA[56]. The most stable helical regions arise from interaction between adjacent sections of the chains, taking the form of hairpin turns, with a minimum of three unbonded nucleotides at the turn. The model must not be

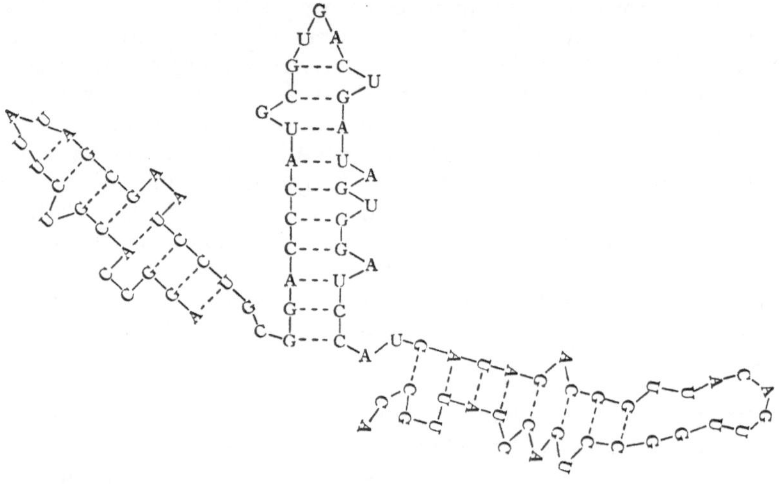

Figure 5.3. Possible[56] base-pairing within singly stranded macromolecular RNA. Helical regions are represented by parallel lines of bases joined by dashed lines, for hydrogen bonds. Bases offset from these lines represent loops outside the helices
(Reproduced by courtesy of the Editor of *Nature, London*)

conceived as a static one, since there is a continual exploration of all possible configurations of the chain.

It would only be possible to accommodate about 20 per cent of the bases in helical regions if the bases were random in sequence and had the proportions actually present in macromolecular RNA, which does not show equality of A with U, or of G with C. However, it has been demonstrated[57] that, in the combination of poly-uridylic acid with the copolymer of adenylic and uridylic acids (*see* page 58), nucleotides not involved in pairing within the helix can be accommodated as loops outside it. On this basis it has been possible to formulate a model for macromolecular RNA in solution in which the RNA chain bends back on itself to form short helices to the extent of involving 50–60 per cent of the nucleotides in short helical regions of about 12 base pairs (*Figure 5.3*). If appropriate non-random sequences occur, higher helical contents could also be explained.

A crystalline form of RNA has been obtained from yeast[58, 59] and has yielded valuable information about the form of at least part of ribosomal RNA. This crystalline material was obtained by alkaline hydrolysis (0·3 N NaOH at 25°C for 5 to 30 minutes, by heating and by other treatments) of a soluble RNA extracted from yeast. The crystallizable material, which had a molecular weight of about 14,000[59, 60], was subsequently shown[59] to be derived entirely from ribosomal RNA and not from transfer S-RNA, as at first thought[58, 59]. The x-ray diffraction patterns for this material were strikingly similar to the *A* type patterns of DNA; and their clarity suggested that most of the molecules had, apart from base sequences, identical and highly regular structures of the DNA type, which implied the usual base-pairing (IX, X) and so complementary sequences. The pitch length observed in this crystallizable RNA varied with humidity from about 28 to 30Å, and the separation of nucleotides along the helix axis direction was 2·5Å, so that the bases were tilted about 20 degrees from the perpendicular to this axis. The significance of these observations is in their making possible the interpretation of the diffuse x-ray diffraction patterns of intact ribosomal and viral RNA. For these diffuse patterns[58] are essentially the same as the crystalline patterns, with only minor variations, and it may be concluded that all types of RNA so far studied contain helical regions which, on average[60], have a configuration essentially similar to that of DNA in the *A* configuration. It is uncertain, from these studies, whether or not the helices are perfectly regular (with respect to mode of base-pairing, and the existence of loops (cf. pages 156–7)) and how much of each RNA molecule is helical, but the general form of the structure in *Figure 5.3* is indicated by these studies.

Further elaboration of the configuration of RNA has been proposed by Spirin[51] who distinguishes a form at low ionic strength and low temperature in which short double-helical regions are stacked alternately with random regions and so folded that the helical axes are perpendicular, and the base rings are parallel, to the axis of one molecule as a whole. This proposal does not alter the basic scheme of *Figure 5.3* except that it involves rather larger random regions between helices than that figure suggests.

The question naturally arises, what is the relation between the configuration of macromolecular RNA in solution and that in the original virus or ribosome? There is an increase of ultra-violet absorption when *E. coli* ribosomes are heated, which is the same[61] as, or at least two-thirds of[42], that observed with RNA isolated from them. Hence that RNA configuration in solution which contains helical regions is essentially retained when the RNA is packed into the ribosome. X-ray diffraction and optical rotatory dispersion studies[62] of ribosomes from other sources also show that such structure as there is in the RNA in these particles is very similar to that of extracted RNA. Although the RNA of TMV in solution has the same configuration as ribosomal RNA, in the virus the single chain of RNA is itself helical along its whole length, and embedded in protein subunits in a way which prohibits hydrogen bonding between bases (Chapter 6; Plate 2). Apparently the protein constrains the RNA into a particular form which is probably essential for the infectivity of the virus. This may not be the situation with all viruses, for certain spherical RNA viruses (bushy stunt, turnip yellow mosaic, and southern bean mosaic) on alkaline hydrolysis exhibit an increase in ultra-violet absorption of a magnitude which suggests that hydrogen-bonded regions exist in the RNA within the virus[61]. Perhaps the explanation is, that in these spherical viruses the RNA is an inner core surrounded by protein and is not a thread constrained to a particular configuration by the protein, as in TMV.

MESSENGER RNA

Studies in the metabolism of nucleic acids in *E. coli* infected by T2-bacteriophage led Volkin and Astrachan[63] to propose the existence of an unstable RNA fraction with base ratios similar to those of the T2-phage (with U instead of T). This proposal was later supported by an examination of the regulatory mechanisms in bacteria[64], further studies on the RNA and ribosomes of *E. coli* after infection by T2-phage[65], and by the use of very short pulses of radioactive precursors in growing uninfected bacterial populations[66]. This fraction of the RNA had the properties expected of 'messenger RNA', the carrier of genetic information from the DNA to the site of protein synthesis in the ribosomes: it constituted only a small fraction of the total RNA ($\ngtr 4$ per cent); it had a rapid turnover; and its base ratios corresponded to those of the infecting phage DNA and were markedly different from those of the host RNA.

Most of this RNA, newly synthesized after T2 infection, is associated with the $70S$ or $100S$ ribosomes of the infected *E. coli* cells, if[65-69] the magnesium ion concentration is above 5×10^{-3}M. It should be noted messenger RNA fractions are present in uninfected as well as in infected cells[66]. A similar RNA fraction, of high turnover rate and with a nucleotide composition like that of DNA, has also been found in yeast cells[70] and in pea seedlings[71].

The molecular characteristics of the messenger RNA are not yet well established: for example, no studies of the helical content have yet been reported. A sedimentation coefficient of 12S and a molecular weight of about $0 \cdot 25 - 0 \cdot 5 \times 10^6$ have been suggested[65] and it is probably heterogeneous, as judged from the skewed shape of the sedimentation diagram. The RNA synthesized by *E. coli* in specific response to infection by T2-bacteriophage had[69] a sedimentation coefficient of about 8S, which is distinct from the usual ribosomal RNA components of about 18S and 25S (*see* page 52). A molecular weight of about $1 - 1 \cdot 5 \times 10^5$ was inferred from the relation of molecular weight to sedimentation coefficient already established[42] for ribosomal RNA.

Messenger RNA of both uninfected and T2-infected cells can be separated[68, 72] on methylated albumin columns into fractions characterized by different sedimentation coefficients thus (following Ishihama and colleagues[72]): I, 8S; II, 10–12S; III, 13S and 19–21S; IV, 23–30S and by their ability to associate with 70S ribosomes (IV combines; II and III do not). It is suggested[72] that fraction IV is possibly the one functionally active in protein synthesis, since it is large enough to code for a protein and combines with ribosomes, whereas II and III may be 'repressor' molecules, as defined by Jacob and Monod[64]. The question of the size of this rapidly-labelled RNA is therefore far from resolved and the possibility remains open that it may be at least as large as the smaller ribosomal RNA (i.e., molecular weight 5×10^5).

ENZYMATIC SYNTHESIS

Although the studies of the configuration of naturally occurring RNA were not at first very encouraging, considerable advances were made in our knowledge of the shape and interactions of the polyribonucleotide chain as a result of the availability of enzymatically synthesized preparations[37, 55, 73]. Ochoa and his colleagues, in an important series of investigations, showed that an enzyme can be isolated from *Azotobacter vinelandii* and other bacteria which, if magnesium is present, catalyses the synthesis of highly polymerized polyribonucleotides from 5'-nucleoside diphosphates, at the same time releasing inorganic phosphate[74]. Other bacteria are also sources of this enzyme. The reaction appears to be $nX-R-P-P \rightleftharpoons (X-R-P)_n + nP$, where R = ribose, P–P = pyrophosphate, P = orthophosphate, X = adenine, hypoxanthine, guanine, uracil or cytosine. Chemical and enzymatic degradation of the polynucleotides $(X–R–P)_n$ has shown that they are made up of 5'-nucleotide units linked by 3',5'-phosphodiester bonds, as in RNA, and that they undergo enzymatic breakdown in a manner similar to that of RNA. The molecular weights vary from 7×10^4 to 3×10^6.

If a single 5'-nucleoside diphosphate was used as the starting material, then a polynucleotide was obtained containing only one type of base; for example, when the starting material was 5'-adenosine diphosphate, the product could be described as polyadenylic acid, or simply poly-A. 'Single' polymers of this type which have in fact been prepared are: poly-A, poly-G (in small amounts), poly-C, poly-U, and poly-I (where I = inosinic acid, containing the base hypoxanthine). If a mixture of 5'-nucleoside diphosphates is used as the starting material, mixed copolymers are obtained in which the different bases are all attached to the same ribose–phosphate backbone; thus, there has been prepared a polymer containing A and U (poly-AU) and a polymer containing A, G, U and C (poly-AGUC). The latter is of particular interest since it corresponds in structure, composition and chain length to naturally occurring RNA. This polymer and poly-AU can be stretched into fibres and give x-ray diffraction patterns very similar to those of natural RNA[75].

The observed similarity of the diffraction patterns shows directly that the synthetic polymers can assume a configuration identical with that of isolated RNA, and this has led to closer investigation of the interactions of the various single polymers. Both poly-C and poly-A in the fibrous state have highly ordered structures and the x-ray pictures of the latter are consistent with a structure in which two helical chains are intertwined, as in DNA, and are linked by hydrogen bonding pairs of adenine rings similar to that occurring in crystals of adenine hydrochloride[76].

Of great interest is the observation[77] that, when the sodium salts of poly-A and poly-U are mixed in equimolar proportions, the two polymers unite to form a two-stranded complex in which both chains are helical and are linked by hydrogen bonds, similar to those in DNA, but with uracil replacing thymine. The x-ray diffraction patterns show that in this case the dimensions are very similar to those of the DNA double helix (namely 3·4 Å internucleotide spacing and 10 nucleotides per turn) thereby proving that poly*ribo*nucleotide chains can take up the same helical configurations as poly*deoxyribo*nucleotides. This may well be of importance in the transfer of 'information', in the form of nucleotide sequences, from DNA to RNA. Of similar importance may be the further observation[78] that the two-stranded (poly-A + poly-U) structure can interact with another poly-U chain, in the presence of magnesium ions, to give a complex in which it is thought that a strand of poly-U is wrapped around the original two strands of (poly-A + poly-U). The possibilities of hydrogen bond formation between bases are sufficiently numerous to allow this, and another similar set of phenomena has been observed in the interaction between poly-A and poly-I. Evidence for the formation of hybrid double helixes containing one strand of a polyribonucleotide and one of a polydeoxyribonucleotide, has been obtained and is discussed later (Chapter 12) in connection with the mechanism for the transfer of genetic information from DNA to RNA.

In view of the interactions between the bases in synthetic polyribonucleotides, there should be many and varied possibilities of interactions between bases in naturally occurring RNA, possessing mixed sequences of nucleotides. The evidence already described indicates that such interactions do indeed determine the configuration of RNA.

REFERENCES

[1] Chargaff, E. and Davidson, J. N., (ed.) *The Nucleic Acids*, Vol. 2, Chapts. 16–19. New York; Academic Press, 1955

[2] Levene, P. A. and Bass, L. W. *Nucleic Acids*. New York; Chemical Catalog Co., 1931

[3] Loring, H. S. In *The Nucleic Acids*, (Ed. Chargaff and Davidson), Vol. 1, p. 191. New York; Academic Press, 1955

[4] Brown, D. M. and Todd, A. R. *Ann. Rev. Biochem.*, 1955, **24**, 311; in *The Nucleic Acids*, (Ed. Chargaff and Davidson), Vol. 1, p. 409. New York; Academic Press, 1955

[5] Jordan, D. O. *The Chemistry of the Nucleic Acids*. London; Butterworths, 1960

[6] Littlefield, J. W. and Dunn, D. B. *Biochem. J.*, 1958, **70**, 642; Smith, J. D. and Dunn, D. B. *Biochem. J.*, 1959, **72**, 294; Cohn, W. E. *J. Biol. Chem.*, 1960, **235**, 1488

[7] Overend, W. G. and Stacey, M. In *The Nucleic Acids*, (Ed. Chargaff and Davidson), Vol. 1, p. 9. New York; Academic Press, 1955

[8] Peacocke, A. R. *Chem. Soc. Special Publ.*, 1957, No. 8, p. 139

[9] Hoagland, M. B., Stephenson, M. L., Scott, J. F., Hecht, L. I. and Zamecnik, P. C. *J. biol. Chem.*, 1958, **231**, 241; Berg, P. and Ofengand, E. J. *Proc. nat. Acad. Sci., Wash.*, 1958, **44**, 78; Berg, P., Bergmann, F. H., Ofengand, E. J. and Dieckmann, M. *J. biol. Chem.*, 1961, **236**, 1726; and other references (Chapter 11); reviewed by Hoagland, M. B. *Proc. IVth Inter. Congr. Biochem.*, Vol. 8, p. 199. Oxford; Pergamon Press, 1959

[10] Tissieres, A. *J. mol. Biol.*, 1959, **1**, 365; Brown, G. L. and Zubay, G. *J. mol. Biol.*, 1960, **2**, 287

RIBONUCLEIC ACID

[11] Ofengand, E. J., Dieckmann, M. and Berg, P., *J. biol. Chem.*, 1961, **236**, 1741; Hecht, L. I, Zamecnik, P. C., Stephenson, M. L. and Scott, J. F. *J. biol. Chem.*, 1958, **233**, 954; Monier, R., Stephenson, M. L. and Zamecnik, P. C. *Biochim. biophys. Acta*, 1960, **43**, 1; Dunn, D. B., Smith, J. D. and Spahr, P. F. *J. mol. Biol.*, 1960, **2**, 113; Fuller, W. *J. mol. Biol.*, 1961, **3**, 175

[12] Apgar, J., Holley, R. W. and Merrill, S. H. *J. Biol. Chem.*, 1962, **237**, 796; Holley, R. W., Apgar, J., Everett, G. A., Madison, J. T., Merrill, S. H. and Zamir, A. *Cold Spr. Harb. Symp. quant. Biol.* 1963, **28**, 117

[13] Karau, W. and Zachau, H. G. *Biochim. biophys. Acta*, 1964, **91**, 549; Thiebe, R. and Zachau, H. G., *ibid.*, 1965, **103**, 568; Melchers, F. and Zachau, H. G., *ibid.*, 1965, **95**, 380; Zachau, H., Dütting, D. and Feldman, H. *Hoppe Seyler's Z. physiol. Chem.*, 1966, **347**, 212

[14] Holley, R. W., Everett, G. A., Madison, J. T. and Zamir, A. *J. Biol. Chem.*, 1965, **240**, 2122; Holley, R. W., Madison, J. T., Zamir, A. *Biochem. Biophys. Res. Comm.* 1964, **17**, 389; Madison, J. T. and Holley, R. W. *ibid.*, 1965, **18**, 153; Melchers, F. and Zachau, H. G. *Biochem. biophys. Acta*, 1964, **91**, 559; Dütting, D. and Zachau, H. G. *ibid.*, 1964, **91**, 573; Melchers, F., Dütting, D. and Zachau, H. G. *ibid.*, 1965, **108**, 182; Biemann, K., Tsunakawa, S., Sonnenbichler, J., Feldmann, H., Dütting, D. and Zachau, H. G. *Angew. Chem. Int. Ed.* 1966, **5**, 590; Hall, R. H., Robins, M. J., Stasiuk, L. and Thedford, R. *J. Amer. chem. Soc.* 1966, **88**, 2614

[15a] McCully, K. S. and Cantoni, G. L. *J. mol. Biol.*, 1962, **5**, 497

[15b] Cantoni, G. L., Ishikura, H., Richards, H. H. and Tanaka, K. *Cold Spr. Harb. Symp. quant. Biol.*, 1963, **28**, 123

[16] Cox, R. A. and Littauer, U. Z. *J. mol. Biol.*, 1960, **2**, 166; Takanami, M., Okamoto, I. and Watanabe, I. *J. mol. Biol.*, 1961, **3**, 476

[17] Spencer, M., Fuller, W., Wilkins, M. H. F. and Brown, G. L. *Nature, Lond.*, 1962, **194**, 1014

[18] Spencer, M. and Poole, F. *J. mol. Biol.*, 1965, **11**, 314

[19] Canellakis, E. S. *Biochim. biophys. Acta*, 1957, **25**, 217; Hecht, L. I., Stephenson, M. L. and Zamecnik, P. C. *Proc. nat. Acad. Sci., Wash.*, 1959, **45**, 505; Preiss, J., Dieckmann, M. and Berg, P., *J. biol. Chem.* 1961, **236**, 1748; Canellakis, E. S. and Hierbert, E. *Proc. nat. Acad. Sci., Wash.*, 1960, **46**, 170

[20] Singer, M. F. and Cantoni, G. L. *Biochim. biophys. Acta.*, 1960, **39**, 182; Zillig, W., Schachtschabel, D. and Krone, W. *Z. Physiol. Chem.* 1960, **318**, 100; Okaka, E. and Osawa, S., *Nature, Lond.*, 1960, **185**, 921

[21] Holley, R. W., Apgar, J., Everett, G. A., Madison, J. T., Marquisce, M., Merrill, S. H., Penswick, J. R. and Zamir, A., *Science, N.Y.*, 1965, **147**, 1462; Holley, R. W. *Scient. Am.*, 1966, **214**, No. 2, 30

[22] Zachau, H. G., Dütting, D. and Feldmann, H. *Angew. Chem., Int. Ed.*, 1966, **5**, 422; *idem. Hoppe Seyler's Z. physiol. Chem.*, 1966, **347**, 212

[23] Zachau, H., Dütting, D., Feldmann, H., Melchers, F. and Karau, W. *Cold Spr. Harb. Symp. quant. Biol.*, 1966, **31**, 417

[24] Ingram, V. M. and Sjöquist, J. A. *Cold Spr. Harb. Symp. quant. Biol.*, 1963, **28**, 133

[25] Bayev, A. A., Venkstern, T. V., Mirzabekov, A. D., Krutilina, A., Li, L. and Axelrod, V. *Biochim. biophys. Acta*, 1965, **108**, 162; Bayev, A. A., Venkstern, T. V., Mirzabekov, A. D., Krutilina, A. I., Axelrod, V.,

REFERENCES

Li, L. and Engelhardt, V. In *Genetic Elements: Properties and Function*, (Ed. D. Shugar) p. 287. London and New York: Academic Press, 1967
26 Madison, J. T., Everett, G. A. and Kung, H. K. *Cold Spr. Harb. Symp. quant. Biol.*, 1966, **31**, 409
27 Lagerkvist, U. and Berg, P. *J. mol. Biol.*, 1962, **5**, 39; Berg, P., Lagerkvist, U. and Dieckmann, M. *J. mol. Biol.*, 1962, **5**, 159
28 Ingram, V. M. and Sjöquist, J. A. *Cold Spr. Harb. Symp. quant. Biol.*, 1963, **28**, 133
29 Crick, F. H. C. *J. mol. Biol.*, 1966, **19**, 548
30 Söll, D., Jones, D. S., Ohtsuka, E., Faulkner, R. D., Lohrmann, R., Hayatsu, H. and Khorana, H. G. *J. mol. Biol.*, 1966, **19**, 556
31 Chargaff, E. In *The Chemical Basis of Heredity*, (Ed. McElroy and Glass), p. 521. Baltimore; Johns Hopkins Press 1957; Magasanik, B. In *The Nucleic Acids*, (Ed. Chargaff and Davidson), p. 373. New York; Academic Press, 1955
32 Elson, D. and Chargaff, E. *Biochim. biophys. Acta*, 1955, **17**, 367
33 Knight, C. A. *J. biol. Chem.*, 1952, **197**, 241
34 Lee, K. Y., Wahl, R. and Barbu, E. *Ann. Inst.*, *Pasteur*, 1956, **91**, 212; Belozersky, A. N. and Spirin, A. S. *Nature, Lond.*, 1958, **182**, 111
35 Schmidt, G., Cubiles, R. and Thannhauser, S. J. *J. cell. comp. Physiol.*, 1951, **38**, Suppl. 1, p. 61
36 Reddi, K. K. *Biochim. biophys. Acta*, 1957, **25**, 528; *Proc. nat. Acad. Sci.*, *Wash.*, 1959, **45**, 293
37 Peacocke, A. R. *Progr. Biophys.*, 1960, **10**, 55
38 Boedtker, H. *Biochim. biophys. Acta*, 1959, **32**, 519; *J. mol. Biol.*, 1960, **2**, 171
39 Kurland, C. G., *J. mol. Biol.*, 1960, **2**, 83
40 Littauer, U. Z. and Eisenberg, H. *Biochim. biophys. Acta*, 1959, **32**, 320
41 Gierer, A. *Z. Naturf.*, 1958, **13b**, 788
42 Hall, B. D. and Doty, P. *J. mol. Biol.*, 1959, **1**, 111
43 Timasheff, S. N., Brown, R. A., Colter, J. S. and Davies, M. *Biochim. biophys. Acta*, 1958, **27**, 662
44 Gierer, A. *Z. Naturf.*, 1956, **11b**, 138
45 Boedtker, H., Moller, W. and Klemperer, E. *Nature, Lond.*, 1962, **194**, 444
46 Möller, W. and Boedtker, H. Acides ribonucleiques et polyphosphates, *Colloq. Intern. Centre Natl. Rech. Sci.*, 1962, **106**, 99
47 Rich, A. and Watson, J. D. *Nature, Lond.*, 1954, **173**, 995
48 Ginoza, W. *Nature, Lond.*, 1958, **181**, 958
49 Gierer, A. *Nature, Lond.*, 1957, **179**, 1297
50 Stent, G. S. and Fuerst, C. R. *J. gen. Physiol.*, 1955, **38**, 441
51 Spirin, A. S. *J. mol. Biol.*, 1960, **2**, 436
52 Jones, A. S., Marsh, G. E., Cox, R. A. and Peacocke, A. R. *Biochim. biophys. Acta*, 1956, **21**, 576; Jones, A. S. and Peacocke, A. R. *Trans. Faraday Soc.*, 1957, **53**, 254
53 Littauer, U. Z. and Eisenberg, H. *Biochim. biophys. Acta*, 1959, **32**, 320; Cox, R. A. and Littauer, U. Z. *Nature, Lond.*, 1959, **184**, 818
54 Hall, B. D. and Doty, P. In *Microsomal Particles and Protein Synthesis*, (Ed. R. B. Roberts), p. 27. Washington; Academy of Science, 1958; Doty, P., Boedtker, H., Fresco, J. R., Haselkorn, R. and Litt, M. *Proc. nat. Acad. Sci., Wash.*, 1959, **45**, 482
55 Doty, P. *Biochem. Soc. Symp.*, 1962, **21**, 8

[56] Fresco, J. R., Alberts, B. M. and Doty, P. *Nature, Lond.*, 1960, **188**, 98
[57] Fresco, J. R. and Alberts, B. M. *Proc. nat. Acad. Sci., Wash.*, 1960, **46**, 311
[58] Spencer, M., Fuller, W., Wilkins, M. H. F. and Brown, G. L. *Nature, Lond.*, 1962, **194**, 1314
[59] Spencer, M. *Cold Spr. Harb. Symp. quant. Biol.*, 1963, **28**, 77
[60] Spencer, M. and Poole, F. *J. mol. Biol.*, 1965, **11**, 314
[61] Schlessinger, D. *J. mol. Biol.*, 1960, **2**, 92
[62] Klug, A., Holmes, K. C. and Finch, J. T. Unpublished observations, 1960; Blake, A. and Peacocke, A. R. *Nature Lond.*, 1965, **208**, 1319
[63] Volkin, E. and Astrachan, L. *Virology*, 1956, **2**, 149; *Biochim. biophys. Acta*, 1958, **29**, 536; Volkin, E., Astrachan, L. and Countryman, J. L. *Virology*, 1958, **6**, 545
[64] Jacob, F. and Monod, J. *J. mol. Biol.*, 1961, **3**, 318
[65] Brenner, S., Jacob, F. and Meselson, M. *Nature, Lond.*, 1961, **190**, 576
[66] Gros, F., Hiatt, H., Gilbert, W., Kurland, C. G., Risebrough, R. W. and Watson, J. D. *Nature, Lond.*, 1961, **190**, 581
[67] Risebrough, R. W., Tissieres, A. and Watson, J. D. *Proc. nat. Acad. Sci., Wash.*, 1962, **48**, 430
[68] Monier, R., Naono, S., Hayes, D., Hayes, F. and Gros, F. *J. mol. Biol.*, 1962, **5**, 311
[69] Nomura, M., Hall, B. D. and Spiegelman, S. *J. mol. Biol.*, 1960, **2**, 306
[70] Ycas, M. and Vincent, W. S. *Proc. nat. Acad. Sci., Wash.*, 1960, **46**, 804
[71] Loening, U. E. *Nature, Lond.*, 1962, **195**, 467
[72] Ishihama, A., Mizuno, N., Takai, M., Otaka, E. and Osawa, S. *J. mol. Biol.*, 1962, **5**, 251
[73] Steiner, R. F. and Beers, R. F. *Polynucleotides—Natural and Synthetic Nucleic Acids*. Amsterdam; Elsevier, 1961
[74] Grunberg-Manago, M., Ortiz, P. J. and Ochoa, S. *Science*, 1955, **122**, 907; Ochoa, S., and Heppel, L. A. In *The Chemical Basis of Heredity* (Ed. McElroy and Glass), p. 615. Baltimore; Johns Hopkins Press, 1957
[75] Rich, A. In *The Chemical Basis of Heredity* (Ed. McElroy and Glass), p. 557. Baltimore; Johns Hopkins Press, 1957
[76] Rich, A., Davies, D. R., Crick, F. H. C. and Watson, J. D. *J. mol. Biol.*, 1961, **3**, 71
[77] Warner, R. C. *Fed. Proc.*, 1956, **15**, 379; Rich, A. and Davies, D. R. *J. Amer. chem. Soc.*, 1956, **78**, 3548
[78] Felsenfeld, G., Davies, D. R. and Rich, A. *J. Amer. chem. Soc.*, 1957 **79**, 2023

CHAPTER 6

NUCLEOPROTEINS

The structure of nucleoproteins[1-5] is of importance because, in the cell nucleus, DNA is almost completely complexed with protein and the transference of genetic information from the DNA to RNA, and hence to the synthetic mechanisms, is likely to be profoundly modified by the presence of the protein. The existence of this complex in a stable form appears to be essential to the structure of the chromosome since agents which dissociate the complex also disrupt the structure of the chromosome. After the early interest in the protein component, attention was largely directed to the DNA, but, in recent years, the nuclear proteins and the nature of their complex with DNA have again been the subject of study. The nature of the interaction of the RNA-protein complex of ribosomes is largely unknown.

DEOXYRIBONUCLEOPROTEINS (DNP)

The double-helical structure has been derived from work on nucleic acid preparations which have been separated from the protein with which they are combined *in vivo*. Both types of nucleic acid seem to occur mainly in association with proteins, although the separated nucleic acids are usually obtained in the form of their salts with sodium or calcium ions, which neutralize the negative charge of the phosphate groups, the only charge on the macromolecule at neutral pH. Nucleic acids have an isoelectric point at an acid pH of about 1–2, so that any molecule which is to neutralize their charge at pH 7 must have an isoelectric point in the alkaline range (that is > 7). In the living organism, the nucleic acids are found associated with proteins and only approximately one-tenth or less of the negative phosphate groups are neutralized by metallic cations[6]. The remainder of the negative charges on the macro-ion must therefore be neutralized by positive charges on the associated protein and it is not surprising that the protein components of DNP contain unusually large proportions of arginine and lysine residues. These latter amino acids (XIV and XV) possess side chains containing amino groups which near to neutrality can become positively charged:

$$
\begin{array}{ccc}
| & & | \\
CO & NH_2{}^+ & CO \\
\diagdown & \diagup\!\!\diagup & \diagdown \\
CH\,(CH_2)_3.C & & CH\,(CH_2)_4.NH_3{}^+ \\
\diagup & \diagdown & \diagup \\
NH & NH_2 & NH \\
| & & | \\
\text{protein} & & \text{protein} \\
\text{chain} & & \text{chain} \\
\text{arginine residue} & & \text{lysine residue} \\
XIV & & XV
\end{array}
$$

The proteins found in conjunction with DNA may be classified as protamines, histones and 'non-histone' or 'residual' protein[1,7]. Although the last of these proteins undoubtedly plays an important part in the intact chromosome, little has yet been discovered of its actual mode of combination with DNA and it will not be further considered here. The DNPs may therefore be subdivided according to the nature of the protein component into deoxyribonucleoprotamines and deoxyribonucleohistones.

Deoxyribonucleoprotamines

The protamines occurring in the DNP of fish sperm contain arginine to the extent of two-thirds of all the amino acid residues and therefore possess a large positive charge at neutral pH. The remaining third of the amino acid residues is non-basic and these residues tend to occur in pairs[3]. Considerable progress has been made in determining the amino acid sequence in some of the components of clupeine (the protamine of herring sperm) and detailed analytical studies of other protamines have been made. Clupeine, however, contains at least six distinguishable components and until these are obtained in sufficient quantities sequence studies will not be conclusive. The amino acid composition of a given protamine is characteristic of the individual fish species, whereas the total DNA composition varies little from one species to another (although, as already suggested, large variations in nucleotide sequence are possible). The protamines are relatively small in size, having molecular weights in the range 4,000–10,000, so that a very large number of such molecules would have to be bound to each DNA molecule to neutralize its charge. The deoxyribonucleoprotamines are of considerable biological interest since it has been proved that the entire nuclear material of the sperm head consists of this complex, which must therefore contain all the paternal genetic information; its structure is also important because it is one of the simplest of the nucleoproteins. The physicochemical evidence[3, 8] suggests that the linkage between the DNA and protamine is predominantly electrostatic. It is dissociated by increasing the concentration of sodium chloride to 2 M and in the nucleoprotamine the amounts of DNA phosphorus ($OP \leqslant O^-$) and protamine arginine

residues $\left[-C \!\!\begin{array}{c} NH_2^+ \\ NH_2 \end{array} \right]$ are almost equal.

X-ray diffraction studies[6] show that in the complex formed between DNA and synthetic poly-L-arginine, the fully extended polyarginine

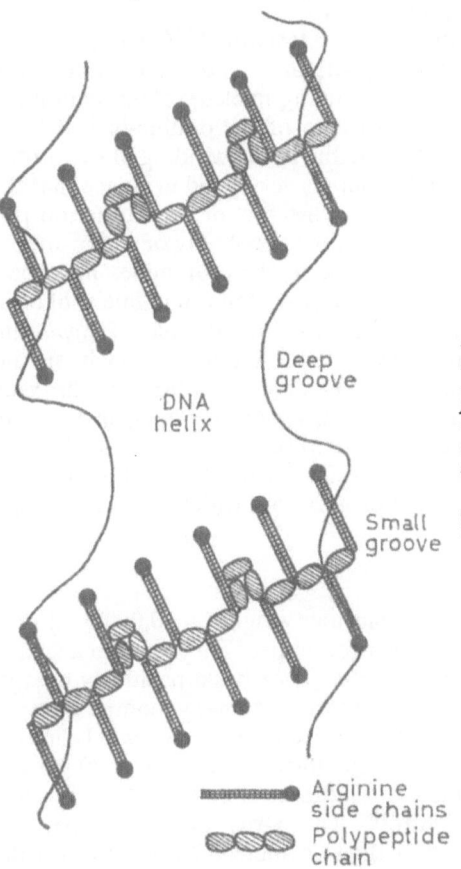

DNA helix

Deep groove

Small groove

Arginine side chains

Polypeptide chain

Figure 6.1. Diagram illustrating how protamine binds to DNA[10]. Black circles represent the phosphate groups of DNA and coincide with the basic ends of the arginine side chains. Non-basic residues are shown as folds in the polypeptide chain

(Reproduced by courtesy of the Editor of *Cold Spring Harbor Symposia on Quantitative Biology*)

5

chain is wound helically round the shallower of the two grooves of the DNA double helix and is anchored to it by the attraction between the positive arginine side chains and the negative phosphate groups (*Figure 6.1*). It was suggested that, since the non-basic amino acid residues occur in pairs in protamines, they could possibly be accommodated as loops in the polyarginine chains, perhaps held by interaction with the base rings of the DNA. These latter interactions could consist of either van der Waals forces or hydrogen bonding, since in the double helix each base pair contains potentially hydrogen bonding groups in the two grooves. This structure now explains the earlier observation of Astbury[9] that the spacing between nucleotides in DNA is equal to the 3·4 Å spacing between amino acid residues in the fully extended β-form of polypeptides.

The x-ray diffraction pictures of complexes of DNA with protamine and with polylysine, of nucleoprotamine extracted from sperm (*Salma trutta*), and of naturally occurring nucleoprotamine in intact fish sperm heads all show the presence of the double-helical DNA structure and show features attributable to the winding of the protein chain around the DNA[10]. Thus, there is very good evidence that this structure occurs *in vivo* and is not an artefact of the extraction procedure. This nucleoprotamine structure is probably of wider importance than in fish species alone, since nucleoprotamines have been found in sperm heads of many animal forms but not in nuclei of other kinds of cell. Even in some sperm cells, for example of *Sepia* and *Loligo*, where the protein component contains a much smaller proportion of basic amino acids than does protamine, the x-ray diffraction picture of the intact sperm heads shows again the main features of the nucleoprotamine structure already described[10]. The proteins in these spermatozoa may almost be classified as histones, which are the other main type of protein with which DNA is found to be associated.

Deoxyribonucleohistones

The histones[11] are larger in molecular weight ($\sim$ 20,000) and more complex in composition than the protamines. They contain a smaller proportion (often less than 30 per cent) of basic positively charged amino acid residues and are mixtures of many components[1, 11]. Certain of these fractions are relatively rich in arginine or in lysine and the complex of DNA with lysine-rich histone, containing 70 per cent of non-basic residues, yields x-ray patterns similar to those given by the nucleoprotamine structure[10]. The x-ray diffraction patterns, however, of unfractionated calf thymus nucleohistone, of intact thymocytes, of unfixed nucleic acid of fowl erythrocytes, and of *Arbacia* sperm heads (all of which contain histone-type proteins) yield a pattern different in certain features from that of nucleoprotamine.

One of the basic problems is the structure of the histone moeity. Studies of the rates of deuteration[12], of the optical properties[4, 12] and of the x-ray diffraction[13] of freshly prepared histones show that it contains the α-helical protein structure to an extent depending on its mode of preparation. This structure gradually changes on storage, even at 5° C, to the extended β form of the denatured state, which is always present in amounts of at least 50 per cent. It is probably in this α-helical form that histone is combined in nucleohistone[13]. From x-ray and other work Zubay and Wilkins conclude that in hydrated nucleohistone, and *in vivo*, hydrated histone gel fills the space between DNA molecules and, although it is associated with the phosphate groups, it is probably not very firmly attached to them[13]. The histone gel has been thought* to be concentrated in regions which form bridges at points approximately 36 Å apart, between parallel DNA molecules[14]. The polypeptide chains are not, to any great extent, preferentially aligned along or perpendicular to the length of the DNA molecules[12, 13].

By water extraction[15, 16] of thymus glands, a deoxyribonucleo-histone has been isolated which has a molecular weight of about 18–20 million and consists of DNA and an approximately equal weight of the smaller histone molecules. This nucleohistone has a reproducible molecular weight and composition and is stable in solution[16]. It contains the DNA in its double-helical form and this is chiefly responsible for its shape in solution[16]. In the complex, there is a wide range of strength of binding of the histone to the DNA, histone rich in arginine being the most strongly attached; combination with the histone stabilizes the core of DNA to denaturation by heat. This water-soluble nucleohistone is a molecular entity and probably forms part of the structure of the chromosome, although the non-histone protein ('residual protein') must also be involved in the complex *in vivo*[17].

There is little knowledge of the molecular interaction in those viruses which contain DNA. It is clear that, in the T2 bacteriophage (*see* Chapter 3) which attacks *E. coli*, double-helical DNA of molecular weight 130×10^6 is enclosed in a polyhedral head (850×700 Å) of protein, to which is attached a narrow tail (1100×150 Å). There is good evidence that the DNA of T2 and other bacteriophages (T5, λ) is all in one large double-helical molecule and this accords with

* But see p. 173.

genetic analysis which shows that all the mutational points of T2 lie on a single unbranched linkage group. It has been suggested that many small viruses may be made up of protein sub-units arranged around a nucleic acid core according to the requirements of symmetry[18]. Careful separation and analysis of the structural components of the even-numbered T bacteriophages certainly show that the protein of the head, sheath and tail consist of sub-units which are characteristic of these structures[19] and evidence is accumulating for this view.

The general conclusion to be drawn from these studies on various types of complex of DNA and protein which occur in biological systems is that the DNA retains its characteristic double-helical structure and acts as the inner core. The various structures afford no obvious model of any highly specific interaction of DNA with the protein component, unless it lies in the suggestion about the looped amino acid side chains and their possible interaction with the base rings. This lack of direct specificity towards protein and the stability of the inner double-helical core may in fact correspond to the genetic and biochemical function of DNA as a carrier of a particular molecular pattern, in the form of a nucleotide sequence, but not as the immediate agent in directing specific protein synthesis, a role which may now be assigned to RNA (Chapters 10 and 11).

RIBONUCLEOPROTEINS (RNP)

It has been known for some time that it is very difficult to remove the last traces of protein from RNA and that a strong interaction between these two species must exist. It is possible that RNA has a specific configuration only when it is attached to protein *in vivo*, and, if this is so, it is not surprising that extracted ribosomal RNA and RNP have yielded little evidence of highly organized structures. The situation has, however, been helped by the existence of the plant viruses[20] which contain only RNA and protein, since these can be isolated and their biological activity may be tested. The RNA in ribosomes appears to have the same molecular configuration as that of isolated RNA, according to x-ray (page 56) and optical rotatory dispersion studies[20a].

The most intensively studied plant virus is TMV and it will alone be described in any detail. It is a rod 3000 Å long, of 150 Å diameter and molecular weight 40×10^6 and contains only 6 per cent by weight of RNA, which when isolated has a molecular weight of the order of $2 \cdot 5 \times 10^6$. This corresponds to the total molecular weight of the RNA

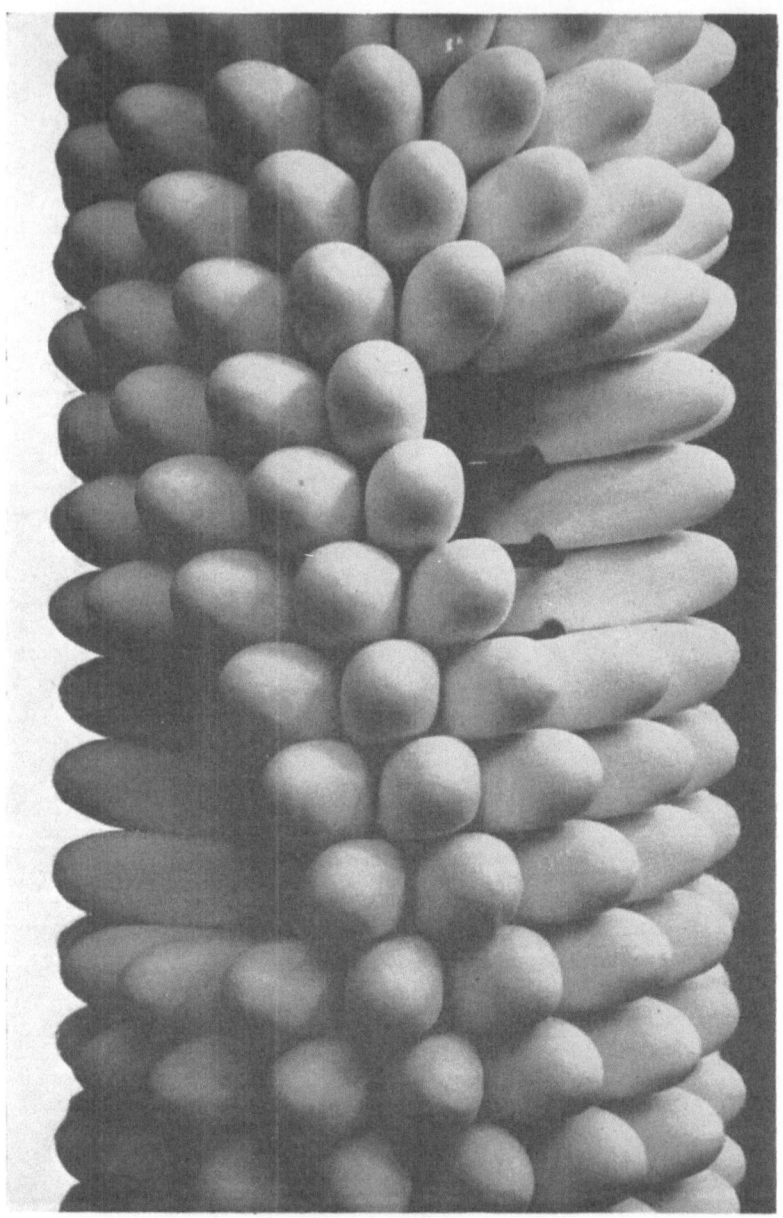

PLATE 2

Model of the structure of tobacco mosaic virus[23]. *The black spiral indicates the position occupied by the single RNA chain between the protein sub-units. Some of the wedges representing these units have been removed to reveal this spiral*

(Reproduced by courtesy of the Editor of *Biological Reviews of the Cambridge Philosophical Society*)

To face p. 69

in the virus and the RNA is present[21] in the form of a single helical strand of diameter 80 Å and a pitch of 23 Å. The RNA lies on the inside of, and is supported by, a helical array of protein units of molecular weight 100,000 (the 'A-protein'), which are each constituted by six smaller sub-units of molecular weight 17,000. There are about 2,100 such smaller sub-units per TMV particle and they probably each correspond to an individual protein chain. An intermediate level of organization appears in the conjunction of eight sets of A-protein (that is, 48 of the smallest sub-units of 17,000 molecular weight) into a unit which has an annular appearance in the electron microscope and consists of three turns of the helical array of sub-units (*Figure 6.2* and Plate 2). Elucidation of the chemical nature of the

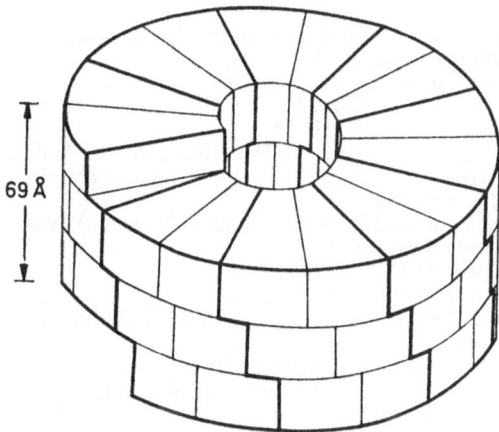

69 Å

Figure 6.2. Schematic model of the annular disc-like unit of TMV. It is made up of 8 molecules of A-protein (the segment within the thick lines), each of which contains 6 protein sub-units, drawn as a truncated wedge[22]

(Reproduced by courtesy of the Editor of *Biochimica et biophysica acta*)

protein chains and the types of chemical bond which hold the various units together should help in understanding the arrangements of other proteins into larger structures. For the present purposes, the important point about the structure is the way in which the predominant protein determines the configuration of the smaller proportion of RNA. The separated protein sub-units can be re-aggregated to give rods of the same diameter as the original TMV, though of variable length, whereas the separated RNA retains its biological activity (Chapter 3) only for a short time after isolation and loses its specific configuration.

The protein (60 per cent) of turnip yellow mosaic virus[24] can exist as a spherical and hollow unit of diameter 200 Å in the absence of RNA and the RNA (40 per cent) when present inside this protein shell has a partially helical structure as it has when isolated (page 52 ff, Chapter 5).

The arrangement of protein in the ribonucleoproteins is probably determined by symmetry considerations, since the protein is made up of a number of sub-units which have to be packed according to geometrical rules. It was on such grounds that Crick and Watson[18] were able to explain the two main forms, spheres and rods, which the viruses assume. Confirmation of their views has been obtained from the x-ray diffraction patterns[25] of the crystals of bushy stunt virus which contains RNA. Icosahedral symmetry, arising from the packing together of 60 sub-units, seems to be the common feature of some of the spherical viruses.

REFERENCES

1 Davison, P. F., Conway, B. E. and Butler, J. A. V. *Progr. Biophys.*, 1954, **4**, 148
2 Magasanik, B. *The Nucleic Acids*, (Ed. Chargaff and Davidson), Vol. 1, p. 373. New York; Academic Press, 1955
3 Felix, K., Fischer, H. and Krekels, A. *Progr. Biophys.*, 1956, **6**, 1
4 Ambrose, E. J. *Progr. Biophys.*, 1956, **6**, 25
5 Peacocke, A. R. *Progr. Biophys.*, 1960, **10**, 55
6 Feughelman, M., Langridge, R., Seeds, W. E., Stokes, A. R., Wilson, H. R., Hooper, C. W., Wilkins, M. H. F., Barclay, R. K. and Hamilton, L. D. *Nature, Lond.*, 1955, **175**, 834
7 Dounce, A. L. *The Nucleic Acids*, (Ed. Chargaff and Davidson), Vol. 2, p. 93. New York; Academic Press, 1955
8 Alexander, P. *Nature, Lond.*, 1952, **169**, 226
9 Astbury, W. T. *Symp. Soc. exp. Biol.*, 1947, **1**, 66
10 Wilkins, M. H. F. *Cold Spr. Harb. Symp. quant. Biol.*, 1956, **21**, 75; *Biochem. Soc. Symp.*, 1956, **14**, 13
11 Cruft, H. J., Hindley, J., Mauritgen, C. M. and Stedman, E. *Nature, Lond.*, 1957, **180**, 1107; Luck, J. M., Cook, H. A., Eldridge, N. T., Haley, M. I., Kuphe, W. D. and Rasmussel, P. S. *Arch. Biochem.*, 1956, **65**, 449; Phillips, D. M. P., *Progr. Biophys.*, 1962, **12**, 211
12 Bradbury, E. M., Price, W. C., Wilkinson, G. R. and Zubay, G. *J. mol. Biol.*, 1962, **4**, 50
13 Zubay, G. and Wilkins, M. H. F. *J. mol. Biol.*, 1962, **4**, 444
14 Wilkins, M. H. F., Zubay, G. and Wilson, H. R. *J. mol. Biol.*, 1959, **1**, 179
15 Doty, P. and Zubay, G. *J. Amer. chem. Soc.*, 1956, **78**, 6707
16 Peacocke, A. R. and Preston, B. N. *Nature, Lond.*, 1961, **192**, 228; Murray, K. and Peacocke, A. R. *Biochem. biophys. Acta*, 1962, **55**, 935; Bayley, P. M., Preston, D. N. and Peacocke, A. R. *Biochim. biophys. Acta*, 1962, **55**, 943; Giannoni, G. and Peacocke, A. R. *Biochim. biophys. Acta*, 1963, **68**, 157; Lee, M. F., Walker, I. O. and Peacocke, A. R. *Biochim. biophys. Acta*, 1963, **72**, 310; Giannoni, G., Peacocke, A. R. and Walker, I. O. *Biochim. biophys. Acta*, 1963, **72**, 469
17 Dounce, A. L. and O'Connell, M. *J. Amer. chem. Soc.*, 1958, **80**, 2013; Kirby, K. S. *Molecular Basis of Neoplasma*, p. 59. Austin: University of Texas Press, 1962; Kirby, K. S. *Progress in experimental Tumour Research*, (Ed. Homberger), Vol. 2, p. 291. Basel; Kargar, 1961; Leveson, J. E. and Peacocke, A. R. *Biochim. biophys. Acta*. 1966, **123**, 329

REFERENCES

[18] Crick, F. H. C. and Watson, D. *Nature, Lond.*, 1956, **177**, 473

[19] Brenner, S., Streisinger, G., Moine, R. W., Champe, S. P., Barnett, L., Berger, S. and Rees, M. W. *J. mol. Biol.*, 1959, **1**, 281

[20] *Cold Spr. Harb. Symp. quant. Biol.*, 1953, **17**; The Nature of Viruses, *Ciba Foundation Symp.* Churchill; London, 1956; Gierer, A. *Progr. Biophys.*, 1960, **10**, 299; *The Viruses*, (Ed. Burnet and Stanley). New York; Academic Press, 1959

[20a] Blake, A. and Peacocke, A. R., *Nature, Lond.*, 1965, **208**, 1319

[21] Klug, A. and Caspar, D. L. D. *Adv. Virus Res.*, 1960, **7**, 225; Klug, A., Holmes, K. C. and Finch, J. T. Unpublished observations, 1960

[22] Klug, A. and Franklin, R. E. *Biochim. biophys. Acta*, 1957, **23**, 199

[23] Franklin, R. E., Caspar, D. L. D. and Klug, A. *Plant Pathology: Problems and Progress*, (Ed. Holton), p. 447. University Press; Wisconsin, 1959

[24] Markham, R. *Disc. Faraday Soc.*, 1951, **11**, 221

[25] Caspar, D. L. D. *Nature, Lond.*, 1956, **177**, 475

CHAPTER 7

CHROMOSOMES

The structure of chromosomes is dependent on the nature and arrangement of the macromolecules and ions of which the chromosome is composed. While the identity of the chemical constituents is well established, the evidence regarding the structural organization of the chromosome is less generally accepted and indeed is often contradictory.

CHEMICAL CONSTITUENTS OF CHROMOSOMES

Evidence regarding the nature of chromosome constituents has been acquired using a number of methods and indicates that chromosomes contain DNA, RNA, basic proteins, acidic proteins and divalent cations, particularly calcium and magnesium.

Cytochemical Methods

The basic principle of all cytochemical methods is to carry out chemical tests specific for certain chemical substances on sections of biological material and then to locate the reaction product by using a microscope. In order to be useful the chemical reaction must be highly specific and the reaction product must remain localized once formed.

Nucleic acids are fairly easily detected and estimated cytochemically by a number of methods. The purine and pyrimidine bases which are present in nucleic acids absorb ultra-violet light and thus can be located using an ultra-violet microscope. Both DNA and RNA have absorption maxima at about 260 mμ and therefore they cannot be distinguished by this method. If, however, one or other of the two nucleic acids has been eliminated from the material by digestion with a specific enzyme (deoxyribonuclease or ribonuclease), the remaining nucleic acid can be located. Specific staining reactions for the detection of DNA and RNA are also available. The first and perhaps easiest of these was the Feulgen reaction[1] which is specific for DNA. In this reaction the purine bases are removed from DNA by acid hydrolysis and the residual apurinic acid gives a strong Schiff aldehyde reaction with bleached basic fuchsin. In addition to showing the site

72

of DNA within the cell this reaction has been used to obtain quantitative estimates of the amount of DNA in cells. Under standard conditions the estimates obtained agree well with those obtained by extraction methods. Basic dyes which react with the phosphoric acid groups of the nucleic acids may be used to locate DNA and RNA. As with the ultra-violet methods, specificity often depends on the selective removal of one of the types of nucleic acid by means of enzymes. There are some methods for the differential staining of DNA and RNA and of these the methyl green–pyronin method[2] is the most widely used. Methyl green stains the DNA while pyronin gives a red colour with RNA.

Cytochemical methods for the detection of protein are on the whole less well developed than those for detecting nucleic acids. Many of the methods do not produce a sufficiently intense staining reaction while with others the staining is too diffuse. By means of the Sakaguchi reaction for protein-bound arginine[3], histones have been located in salivary gland[4] and other chromosomes[5]. The Millon reaction has also been widely used for detecting protein in tissue. This reaction results in the formation of a coloured compound when the hydroxyl groups of tyrosine and tryptophan react with mercuric ions and nitrous acid and has demonstrated that acidic proteins are also present in chromosomes.

Direct Chemical Analysis

Two main methods have been used to obtain isolated nuclei from various tissues. One method involves the homogenization of the tissue in citric acid–sucrose solution followed by differential centrifugation[6] whereas in the other method the nuclei are isolated in density gradients formed with non-aqueous solvents[7]. When such nuclei are disrupted and centrifuged a second time, fine thread-like filaments can be obtained[8]. There have been doubts[9] that these fibrils were intermitotic chromosomes but Mirsky[10] concluded that they were because: (1) they are differentiated along their length in a manner similar to chromosomes; (2) many of them are obviously double; and (3) preparations from three different tissues yielded structures with the same individual characteristics, including one structure identified as a chromosome with the nucleolus attached to it. Moreover Feulgen-negative regions, corresponding to centromeres, have been clearly shown in many of these isolated filaments.

Chemical analysis of the isolated fibres indicates that 90 per cent of the total weight of the chromosome of somatic cells is DNA and histone together, with a little less DNA than histone. In sperm cells[10] the histone is replaced by protamine (*see* Chapter 6). Extraction of isolated chromosomes with 1 M sodium chloride separates the nucleohistone (90 per cent) from a residue (10 per cent) of RNA and a 'residual' protein, or proteins, which is acidic and contains the amino acid tryptophan, which is present only in trace amounts in histones. The RNA forms about 10 per cent of the residue (that is, about 1 per cent of the original chromosome).

Isotopic Methods

The most widely used isotopic method is autoradiography. In this, cells or tissue are incubated with a compound containing radioactive atoms and squash preparations are then made. These squash preparations are stained in the usual manner. Additionally the squashes are covered with emulsion which is sensitive to the β-particles emitted on the disintegration of the radioactive atoms. When the preparation is developed and fixed, dark spots are produced in the emulsion where it has been penetrated by the β-particles. The tracks of these particles are often very short (about 1 μ in the case of tritium) and so it is fairly easy to see in which part of the cell the atom is located by comparing the pattern of spots in the emulsion with the normal cytological preparation over which it is superimposed.

Using tritiated thymidine, which is incorporated exclusively into DNA, it has been shown that the chromosomes contain DNA[11]. In experiments with labelled uridine, which is incorporated into RNA but not DNA, Sirlin confirmed that the chromosomes also contain RNA[12]. He also proved that protein is a constituent of chromosomes, by using ^{14}C-labelled phenylalanine.

Miescher's analysis of fish sperm showed that they contained considerable amounts of calcium and there are now numerous reports[13] of the occurrence of divalent metal ions in nuclei and chromosomes. Mazia[14] suggested that calcium and magnesium ions may be involved in linking macromolecules together in chromosomes, but this is by no means certain. A direct demonstration that calcium occurs in chromosomes has been obtained in Lilium[15] and in Habrobracon[16] by means of autoradiography but the technique does not permit the precise localization of the metals in the chromosomes. There is some other evidence that the cations are bound to the nucleic acid phosphates and that they help to stabilize the structure.

Enzymatic Methods

Two groups of enzymes, proteases and nucleases, have provided information regarding the composition and structure of chromosomes.

If a suspension of isolated chromosomes is treated with trypsin, the viscosity of the fluid increases indicating that DNA is going into solution and microscopic examination shows that the chromosomes have disintegrated[10]. The effect of trypsin demonstrates that protein is present in the chromosome and that it is necessary for the maintenance of chromosome structure. Pepsin, another proteolytic enzyme, while causing changes has a much less drastic effect on chromosome structure.

When isolated chromosomes free from histone are treated with deoxyribonuclease, marked changes occur in their microscopic appearance. The DNA goes into solution and an insoluble residue of acidic protein is left. Treatment of chromosomes with ribonuclease reduces the staining intensity when dyes such as pyronin or toluidine blue are used. This has been interpreted as evidence that RNA is present in chromosomes.

In the newt, *Triturus cristatus*, large diplotene chromosomes may be isolated from oöcytes and maintained for several days in buffer. These 'lampbrush chromosomes' are sufficiently large for the action of enzymes on the chromosomes to be studied microscopically. Evidence obtained from this material supports the conclusion that chromosomes contain protein, RNA and DNA[17].

STRUCTURAL ORGANIZATION OF CHROMOSOMES

There are two basic problems of chromosome structure which have not yet been solved: firstly, the number of strands present; and secondly, the linear differentiation which exists along the chromosome or its constituent strands.

Lateral Organization

It is widely accepted that, at all stages of the mitotic and meiotic cycles, chromosomes are multistranded. This view is supported by microscopic observation and also by experiments yielding, less directly, information about the subdivision of chromosomes. There is, however, a body of apparently equally valid evidence which suggests that, prior to duplication during interphase, the chromosome is indivisible in the lateral sense.

Kaufmann[18] was probably the first to suggest that at anaphase the chromosome is two-stranded and later Nebel[19] and Kuwada[19] proposed that each of these strands was again split giving a four-stranded structure at anaphase. At prophase such a chromosome would consist of two chromatids each containing two strands. These conclusions were based on the observation of the unravelled free ends of chromosomes which were treated with various agents before being fixed and stained and it has been argued that the unravelled ends are an artefact. More recently Grell[20] observed that while in early prophase the chromosomes of some tissues of the mosquito appear to be single. Later in the same prophase each chromosome may be resolved into 16 or even 32 separate strands. The multistranded nature of the chromosome at either the first or second anaphase of meiosis in Podisma, Llareiella, and Trillium has been proposed[21]. In maize, B-chromosomes occasionally divide at both first and second meiotic divisions[22] which implies that at times these chromosomes are multistranded. Manton[23], using ultra-violet photography, obtained good evidence that in the fern Todea the anaphase chromosomes have four strands. By inhibiting nuclear division in sea urchin eggs with mercaptoethanol, Mazia[24] provided evidence that the prophase chromosome contains at least four strands. In cells whose division is blocked by mercaptoethanol the centriole continues its development and may divide to form a four-poled system. If the inhibitor is then removed, the egg immediately divides into four, each cell receiving a full chromosome complement. DNA synthesis does not occur while the cell is blocked so that each of the four daughter nuclei contains only half the normal amount of DNA and presumably half the normal number of strands per chromosome.

Electron microscope studies have confirmed and extended the multistranded nature of the chromosome which was indicated by the light microscope. The higher resolution of electron microscopy has shown that the prophase chromosome has a fibrillar structure resembling a multistranded cable[25]. Ris[26] studied squash preparations of lampbrush chromosomes from various amphibia, pachytene chromosomes from a number of insects and plants and leptotene chromosomes from Lilium. He concluded that all these types of prophase chromosome consist of fibrils about 500 Å in diameter. The use of ultra-thin sections instead of squash preparations improved the resolution obtainable with the electron microscope and revealed that the 500 Å fibrils consisted of two strands each of which is 200–250 Å wide[26]. Further improvements in technique have resulted in the demonstration of 100 Å fibres[27] in ultra-thin sections of many chromosomes and it has been suggested that this structure is the basic unit of the chromosome[25]. Fibrils of this diameter can be obtained by extracting calf thymus nuclei with a solution of sodium chloride (0·9 per cent) and Versene at pH 6·5. If, however, the extraction is carried out at pH 8 electron micrographs show that the solution contains fibres 25–40 Å thick or pairs of these fibres. These 40 Å filaments are thought to correspond to a nucleohistone fibril and to consist of a DNA double helix and associated histone. The generally accepted view which has emerged from electron microscopic studies of higher organisms is that before DNA synthesis, the chromosome is composed of eight strands each 200–250 Å wide which in turn each consist of two fibrils about 100 Å in diameter. Each 100 Å thread is regarded as composed of two DNA double helices with their associated protein (*Figure 7.1*).

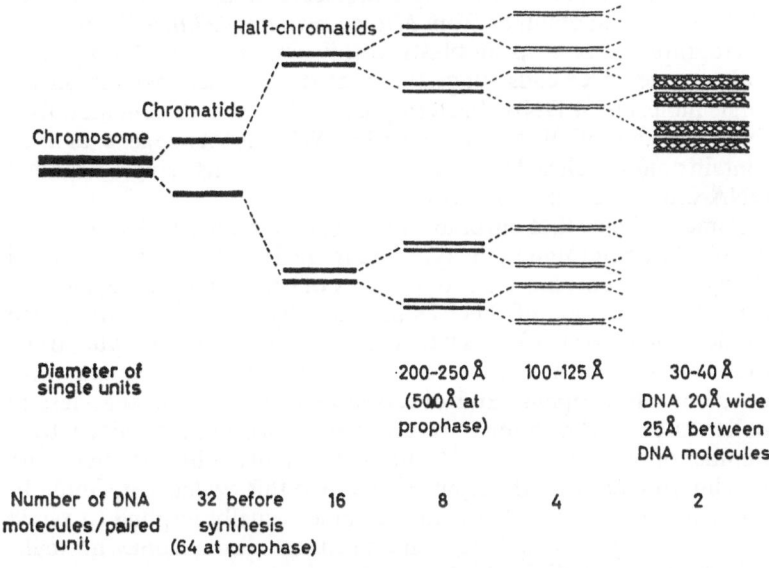

Diameter of single units		200–250 Å (500 Å at prophase)	100–125 Å	30–40 Å DNA 20Å wide 25Å between DNA molecules

Number of DNA molecules/paired unit	32 before synthesis (64 at prophase)	16	8	4	2

Figure 7.1. A diagrammatic representation[16] of a multistranded chromosome showing a progression of strands in pairs to the single nucleoprotein unit

(Reproduced by courtesy of the Brookhaven National Laboratory)

In addition to the work by electron microscopists on the structure of chromosomes in higher organisms some fairly extensive investigations have been carried out on lower forms. Marshak[28] obtained evidence that the bacterial chromosome contained two strands each about 500 Å in diameter and each divided in two. More recently Kellenberger[29] examined *E. coli* and found the chromosome to be multistranded, composed of fibrillae 20–60 Å thick. In the bacterial nuclear structure, DNA fibres exist free of histones and other proteins so that the fibres isolated may be DNA double helices. So it has been suggested on the basis of Kellenberger's results that the chromosome of *E. coli* is a bundle of eight DNA molecules[16].

That this interpretation is almost certainly wrong has been demonstrated by the autoradiographic studies of Cairns[30] which he made on *E. coli*, as described in Chapter 4 (*see* page 36). From the results it appears probable that the chromosome of *E. coli* is one continuous double-helical molecule of DNA and that this normally exists as a circle (*see* Addendum, page 171). Other, less conclusive, evidence for the presence of only a single molecule of DNA in a bacterial nucleus has been obtained from *Micrococcus lysodeikticus*[31]. Electron micrographs of lysed protoplasts of this organism reveal long fibres with very few free ends suggesting that the nucleus contains only a single molecule of DNA. Bacteriophages also seem to contain only a single molecule of DNA[32] (*see* page 35). While most phages examined contain double-helical DNA at least one, ϕX-174, has a single-stranded DNA which is circular in form[33].

Some evidence that chromosomes may be multistranded has been obtained by determining the type and frequency of induced structural changes in the chromosomes of progeny of irradiated individuals. An early demonstration of this was provided by Nebel[34] in x-irradiated Tradescantia. His work was later confirmed by Swanson[35] who, using the same organism, demonstrated half-chromatid exchanges produced at late prophase and observed at metaphase or anaphase in pollen tube nuclei. Similar results have since been obtained by a number of other workers[21, 36], although not all are in agreement with this interpretation of the results[37]. Despite this evidence it should be pointed out that interphase chromosomes usually respond to x-rays as though they were undivided and prophase chromosomes normally behave as though divided into two strands.

Spectrophotometric studies of the DNA-Feulgen reaction have led a number of workers to conclude that chromosomes are multistranded. The chromosome number, chromosome size and DNA content of nuclei in related species of insects have been compared[38] and it was found that, although the chromosome number differed by a factor of two, the amount of DNA per nucleus remained relatively constant. It was concluded that in Thyanta and Banasa the doubling of the chromosome number in some species was accompanied by a reduction of the number of strands in each chromosome thus maintaining the DNA content of the nucleus. A correlation between the amount of DNA in somatic nuclei and the width of the loops in lampbrush chromosomes has been pointed out by Ris[26] who suggested that both increases are related to an increase in the number of strands present in the chromosomes of the different species. The delay in the appearance of mutations following the use of some chemical mutagens has been employed as evidence for the multistranded nature of chromosomes. Such delays may involve a few or many cell generations and have been interpreted as being due to the sorting out, through successive mitoses, of a recessive mutation which affected only one sub-unit until the mutation is present in all strands of the chromosome. Mutations of this type were demonstrated first in Drosophila[39] but have since been studied more extensively in *E. coli*[40]. In view of Cairns' demonstration that the chromosome in this organism is in fact all in one double helix (*see* page 36 and Addendum, page 171) it would appear that, in this organism at least, sorting out of sub-units could not explain the delay in the appearance of mutations.

The study of the lampbrush chromosomes found at the diplotene stage in amphibian oöcytes has proved controversial, but it now seems certain that these chromosomes are not multistranded and have only one DNA double helix per chromatid. Lampbrush chromosomes are remarkably long and produce pairs of looped projections from the main axis of the chromosome. At later stages in division the chromosomes shorten considerably and the loops are retracted. Callan and McGregor[41] showed that treatment of isolated lampbrush chromosomes of newts with deoxyribonuclease severed the chromosome either in the axis of the loop or in the main axis of the chromosome. These chromosomes contain both RNA and protein in addition to DNA but neither ribonucleases nor proteases break the main or the loop axis, which implies that DNA provides the structural continuity of the chromosome. From measurements of the rate at which the loops are broken by deoxyribonuclease, Gall[42] calculated that the axis of the *loops* is one double helix of DNA. Similar calculations show that the structural basis of the main axis of the chromosome is a pair of DNA double helices, which confirms that the chromosome of this organism contains one double helix of DNA before duplication. Treatment of such chromosomes with concentrated potassium chloride removes much of the RNA and protein from the loops and leaves a fine fibril about 40–50 Å in diameter[43], presumably composed mainly of DNA.

Some conclusions regarding the number of chromosomal sub-units may be derived from autoradiographic studies of the duplication of chromosomes labelled with tritiated thymidine. The first experiments of this nature were carried out with the broad bean by Taylor, Woods and Hughes[44]. In these experiments bean seedlings were grown for a time in a solution containing ³H-thymidine, and then in a medium containing no thymidine. At various intervals plants were removed and fixed and the root tips were examined cytologically and by autoradiography. After one cycle of duplication in the presence of ³H-thymidine, both chromosomes descended from an original unlabelled chromosome were labelled (*Figure 7.2*) and sister chromatids appeared to be equally labelled. Following a further mitotic cycle in the absence of thymidine each labelled chromosome produced one labelled and one unlabelled descendant. In a few cells which had gone through one

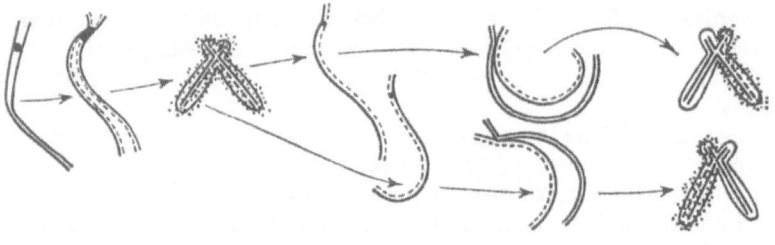

| Duplication with labelled thymidine | 1st metaphase after labelling | Duplication without labelled thymidine | 2nd metaphase after labelling |

Figure 7.2. Diagrammatic representation[64] of the distribution of sub-units of DNA during chromosome duplication. Although the sub-units are not resolved by microscopic examination, they are represented and distributed as indicated by the observed pattern of labelling. Broken lines represent labelled sub-units and unbroken lines represent unlabelled sub-units. The dots represent grains in the autoradiograms

(Reproduced by courtesy of the Academic Press, New York)

more division, half of the chromosomes were unlabelled while in the other half each chromosome had one labelled and one unlabelled chromatid. These results are those expected if the interphase chromosome before duplication contains two components each of which remains intact and functions as a template for DNA synthesis during chromosome duplication which is therefore semi-conservative, as depicted in III of *Figure 8.2*. Colchicine was used to prevent spindle formation during these experiments and LaCour and Pelc[45] criticized the results obtained on this ground, but their criticisms have been satisfactorily countered[46]. Further evidence for such semi-conservative distribution of tritium-labelled DNA in chromosomes has now been obtained for a number of plants and animals[47]. The segregation of tritium-labelled DNA in *E. coli* has been studied and the results show that two large DNA-containing units segregate regularly at cell division. This pattern remains constant for many generations and is consistent with the view that the chromosome is composed of two units each replicating in a semi-conservative fashion. While these results cannot determine the question, it does not seem likely that the individual duplicating units of the chromosomes are each an intact multistranded structure.

Linear Organization

Microscopic studies show that the chromosome is linearly differentiated into lightly and more darkly staining regions. These more darkly staining segments, known as chromomeres, represent at least four types of structure[48]. Some workers identified the chromomeres as the actual genes but this view is not generally held and has been contradicted by the demonstration that, in maize, genes may be located in the lightly staining region between chromomeres[49].

It is now widely agreed that chromomeres are tightly coiled regions of the chromosome and that in the intervening regions the chromosomal fibre is much less coiled. This is confirmed by studies on various plants and animals in which it has been shown that the chromomeres may be resolved into coils by a variety of treatments[48]. Electron micrographs show clearly that the chromomeres represent segments of more intense coiling. In addition such pictures indicate that the fibres in the chromomere are continuous with those in other regions.

Evidence that the chromosome fibril might be formed by an end-to-end aggregation of rod-like particles has been produced by several workers[50]. Mazia obtained nucleoprotein particles, 4000 Å long, from sea urchin sperm under conditions which remove calcium and magnesium from the chromosome and he concluded that divalent cations linked these particles together in the chromosome. Ambrose[50], on the other hand, treated salivary gland chromosomes with agents known to break hydrogen bonds and interpreted his results as indicating that the chromosome consisted of rods linked together end-wise by hydrogen bonds. Specific objections to the role of divalent cations proposed by Mazia have been made[51]. The function of divalent cations in chromosome structure has been reviewed by Steffensen[52] without any definite conclusions being reached as to their importance in determining linear organization of chromosomes.

Several models for chromosomes have been suggested in recent years but none is entirely satisfactory. The basic consideration in chromosome models is whether the main axis of the chromosome is DNA, protein or both. Models in which protein forms the main axis

6   81

have DNA projecting as side chains from the protein[53]. However the feasibility of such models may be questioned on several grounds. None of the genetic results obtained from crosses involving markers belonging to the same or different genes support the idea of a branched chromosome and indeed most results disagree with such a structure[54]. When cells are treated with ultra-violet light the action spectrum for production of chromosome breaks is almost identical with the absorption spectrum of nucleic acids[55]. Tests show no detectable transfer of energy from the nucleic acids[56], which suggests that the chromosome breaks result from a break in the DNA itself and favours a model of the chromosome in which the DNA is involved in its continuity. Callan and McGregor[41] and Gall[42] proved beyond doubt that DNA is involved in maintaining the linear integrity of the chromosome by showing that deoxyribonuclease, but neither ribonuclease nor proteolytic enzymes, can sever the loops and main axis of lampbrush chromosomes. While these experiments show that DNA is necessary to maintain the structural integrity of the chromosome they do not rule out the possibility of other substances alternating with DNA along the axis. Extensive peptide or RNA regions are unlikely since neither ribonuclease nor proteases break the axis. Cairns found no evidence for linkers in his autoradiographic work with *E. coli* but his experiments did not rule out the possibility of short linker segments being present.

While a chromosome model based on a single DNA double helix extending throughout the chromosome fits many of the facts available it seems to lack certain necessary characteristics. Most DNA preparations have a molecular weight of about $5–12 \times 10^6$, which is a very small part of the total DNA of a chromosome. While a certain amount of degradation may occur during isolation, it is also possible that chromosomes could contain molecules with this average weight and that these molecules are linked together to form the chromosome axis. Other evidence favouring the presence of linking units is based on the supposed method of DNA replication. The results of Meselson and Stahl[57] and Cairns[30] require that the two strands in a double helix unwind before replication. In many organisms the chromosome contains enough DNA to correspond to a continuous double helix several metres long and, while it has been calculated[58] that such a helix could unwind and replicate without breaking, it would seem to be more advantageous to the organism if the chromosome contained a number of sites around which it could rotate freely during replication[59]. The work of Kellenberger[60] suggests that phage DNA is folded into a compact structure before the head membrane is formed round it.

Under these circumstances it is reasonable to suppose that some at least of the folding is due to structural features in the DNA molecule. Regions concerned solely or primarily with the specification of the tertiary structure of DNA in bacteriophage have been postulated by Mahler and Fraser[61] to explain some results obtained in experiments with phages T2 and T4.

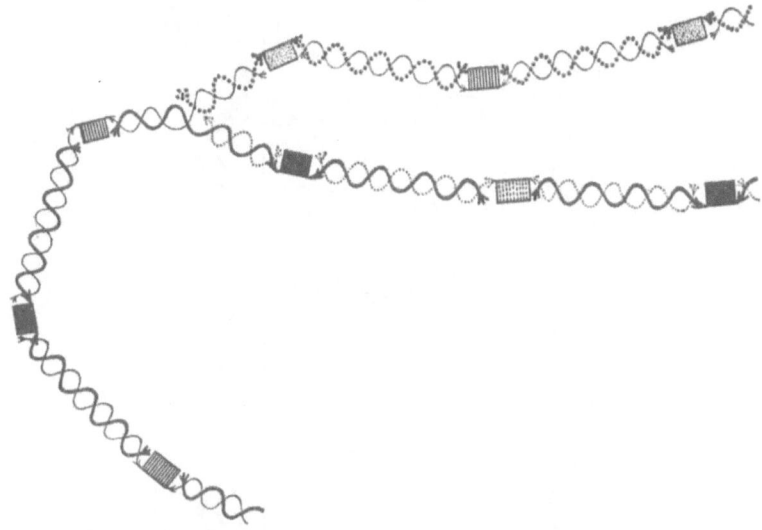

Figure 7.3. Diagrammatic figure[54] of a part of a chromosome model during DNA duplication in interphase. The DNA duplicates and the connecting units (boxes) constitute the alternating elements of the one-dimensional chromosomal thread. During duplication, as a rule, the connecting units stay together with those DNA chains to which they are attached. These coherent systems are indicated in the unduplicated part of the Figure (left) by heavy lines and black boxes and by light lines and striped boxes respectively. In the duplicated part (right) the new systems of DNA chains and connecting units are dotted

(Reproduced by courtesy of the Editor of *Cold Spring Harbor Symposia on Quantitative Biology*)

Connecting units have not yet been identified by analysis of DNA but this is not surprising if they are of low molecular weight and separate DNA segments with molecular weights of several million. The work of Callan's group on lampbrush chromosomes appears to rule out peptides as linker units but there are two observations which may give some indication of the chemical nature of the linkers. Analysis of the DNA from phage T2 indicates[62] the presence of about 500 molecules of 6-methylaminopurine per particle of DNA. These residues cannot form normal hydrogen bonds with thymine and thus might cause an alteration in the structure of the molecule. The other observation which may be relevant is that sperm DNA contains about

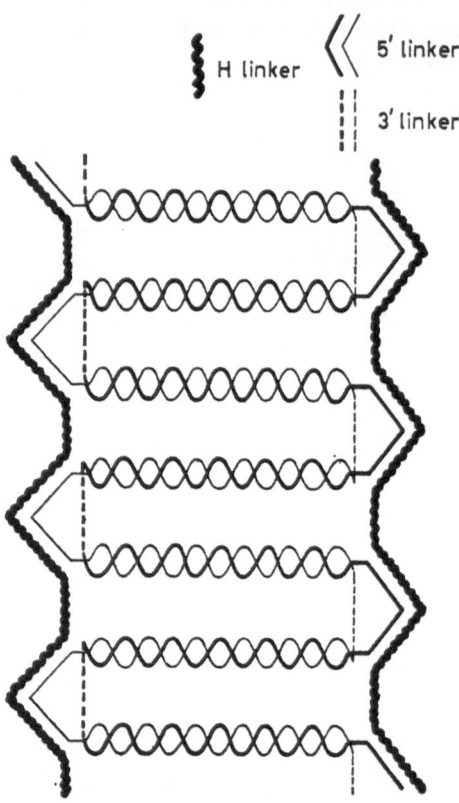

The chromosome structure proposed by Taylor[64]

Figure 7.4. Before or during replication the chromosome is assumed to be stabilized by the addition of H linkers to form a double axis, each of which with its attached DNA chains will become a chromatid by the time replication is complete. The stabilizing material is represented as a polymer attached to alternate 5' linkers. Although shown as a continuous chain for simplicity in illustration, it probably would consist of many small sub-units attached in some way as yet unknown

(Reproduced by courtesy of the Academic Press, New York)

one phosphoserine residue per 1,000 nucleotides[63]. Taylor[64] suggested that the nucleoside ends of the DNA chains in adjacent molecules may be linked through pairs of these phosphoserine molecules.

A specific chromosome model, whose essential feature is a series of double helices linked in tandem by connecting units, was proposed by Freese[54] (*Figure 7.3*). Each end of a DNA molecule is attached to the linker by only one chain the other half of the DNA helix being unattached to this linker. An additional assumption is that all bonds between DNA and linker units are of the same type involving only the nucleoside or the nucleotide end of the polynucleotide chains. In *Figure 7.3* this is indicated by the arrows showing direction in the polynucleotides. This model is in agreement with the results of Taylor[65] on chromosome duplication and sister strand exchange. Such a model is also consistent with the requirements of recombination and of mutation. While the suggested model adequately describes the chromosome during interphase some modification is necessary to accommodate the much condensed coiled structure seen at other stages of division. Freese suggested that the development of this tertiary structure might involve the attachment of the connecting units to each other at regular intervals along the thread forming a structure similar to that in *Figure 7.4*.

Various modifications of Freese's model have been suggested and these have been carried farthest by Taylor[64]. The structure proposed by Taylor (*Figure 7.4*.) differs from that of Freese in two main respects. Two types of link, which join adjacent double helices, are suggested. One joins the 3' carbons of terminal nucleosides in adjacent molecules and the other joins the 5' carbons of the terminal nucleotide in the complementary chains of the same molecules. The other difference in the two models is that Taylor postulates a third connecting unit, the 'H' linker, as essential for the formation of the tertiary structure during metaphase and anaphase whereas in Freese's model this structure is dependent on the lateral linking of connecting units already present.

REFERENCES

[1] Feulgen, R. and Rossenbeck, H. *Hoppe-Seyler's Z. Physiol. Chem.*, 1924, **135**, 203

[2] Brachet, J. *Quart. J. micr. Sci.*, 1953, **44**, 1; Kurnick, N. B. *J. gen. Physiol.*, 1950, **33**, 243; Kurnick, N. B. and Mirsky, A. E. *J. gen. Physiol.*, 1950, **33**, 265

[3] Thomas, L. E. *Stain Tech.*, 1950, **25**, 143

[4] Serra, J. A. and Queiroz-Lopes, A. *Naturwissenschaften*, 1944, **32**, 47

[5] Alfert, M. and Geschwind, T. L. *Proc. nat. Acad. Sci., Wash.*, 1953, **39**, 991

[6] Claude, A. and Potter, J. S. *J. exp. Med.*, 1943, **77**, 345

[7] Allfrey, V. G., Stern, H., Mirsky, A. E. and Saetren, H. *J. gen. physiol.*, 1952, **35**, 529

[8] Mirsky, A. E. and Ris, H. *J. gen. Physiol.*, 1947, **31**, 1

[9] Lamb, W. G. P. *Exp. Cell Res.*, 1950, **1**, 571

[10] Mirsky, A. E. In *Genetics in the Twentieth Century*, (Ed. Dunn), p. 127 New York; Macmillan Co. 1951

[11] Taylor, J. H., Woods, P. S. and Hughes, W. L. *Proc. nat. Acad. Sci., Wash.*, 1957, **43**, 122

[12] Sirlin, J. L. *Exp. Cell Res.*, 1960, **19**, 177

[13] Milovidov, P. F. *Protoplasmalogia Monographien No. 20.* 1949

[14] Mazia, D. *Proc. nat. Acad. Sci., Wash.*, 1954, **40**, 521

[15] Steffensen, D. M. and Bergeron, J. A. *J. biophys. biochem. Cytol.*, 1959, **6**, 339

[16] Steffensen, D. M. In *Structure and Function of Genetic Elements*, p. 103. New York; Brookhaven Nat. Lab. 1959

[17] Callan, H. G. and Lloyd, L. *Phil. Trans. Roy. Soc.*, 1960, **243B**, 135

[18] Kauffmann, B. P. *Amer. Nat.*, 1931, **65**, 280

[19] Kuwada, Y. *Cytologia, Tokyo*, 1939, **10**, 213; Nebel, B. R. *Z. Zellforsch.*, 1932, **16**, 251; Nebel, B. R. *Bot. Rev.*, 1939, **8**, 563

[20] Grell, S. M. *Genetics*, 1946, **31**, 60

[21] Makino, S. *J. Fac. Sci., Hokkaido Univ.*, 1936, **5**, 29; Hughes-Schrader, S. *Biol. Bull.*, 1940, **78**, 312; Wilson, G. B., Sparrow, A. H. and Pond, V. *Amer. J. Bot.*, 1959, **46**, 309

[22] Blackwood, M. *Heredity*, 1956, **10**, 353

[23] Manton, I. *Amer. J. Bot.*, 1945, **32**, 342

[24] Mazia, D. *Ann. N.Y. Acad. Sci.*, 1960, **90**, 455

[25] Ris, H. and Chandler, B. L. *Cold Spr. Harb. Symp. quant. Biol.*, 1963, **28**, 1

[26] Ris, H. In *The Chemical Basis of Heredity*, (Ed. McElroy and Glass) p. 23, Baltimore; Johns Hopkins Press, 1957

[27] DeRobertis, E. *J. biophys. biochem. Cytol.*, 1956, **2**, 785

[28] Marshak, A. *Exp. Cell Res.*, 1951, **2**, 243

[29] Kellenberger, E. In *The Interpretation of Ultra-structure*, (Ed. Harris) p. 223, New York; Academic Press, 1962; Kellenberger, E., Ryter, A. and Séchaud, J. *J. biophys. biochem. Cytol.* 1958, **4**, 671; Ryter, A., Kellenberger, E., Birch-Anderson, A. and Maaløe, O. *Z. Naturf.*, 1958, **136**, 597

[30] Cairns, J. *J. mol. Biol.*, 1962, **4**, 407

[31] Kleinschmidt, A., Long, D. and Zahn, R. K. *Z. Naturf.*, 1961, **166**, 730; *Proc. Eur. Reg. Conf. El. Micr.*, 1960, **2**, 690

[32] Thomas, C. A. In *Molecular Genetics Part 1*, (Ed. Taylor) p. 113, New York; Academic Press, 1963

[33] Fiers, W. and Sinsheimer, R. L. *J. mol. Biol.*, 1962, **5**, 408, 420, 424

[34] Nebel, B. R. *Amer. J. Bot.*, 1937, **24**, 365

[35] Swanson, C. P. *Proc. nat. Acad. Sci., Wash.*, 1947, **33**, 229

[36] Crouse, H. V. *Chromosoma*, 1961, **12**, 190; Sax, K. and King, E. D. *Proc. nat. Acad. Sci., Wash.*, 1955, **41**, 150

[37] Ostergren, G. and Wakonig, T. *Bot. Nat.*, 1954, **4**, 357

[38] Schrader, F. and Hughes-Schrader, S. *Chromosoma*, 1956, **7**, 469 and *Chromosoma*, 1958, **9**, 193

[39] Auerbach, C. *Cold Spr. Harb. Symp. quant. Biol.*, 1951, **16**, 199

[40] Demerec, M. *Symp. Soc. Exp. Biol.*, 1953, 7, 43 and *Amer. Nat.*, 1955, **89**, 5

[41] Callan, H. G. and MacGregor, H. C. *Nature, Lond.*, 1958, **181**, 479

[42] Gall, J. G. *Nature, Lond.*, 1963, **198**, 36

[43] Miller, O. L. Quoted in *Molecular Genetics Part 1*, (Ed. Taylor) p. 69, New York; Academic Press, 1961

REFERENCES

[44]Taylor, J. H., Woods, P. S. and Hughes, W. L. *Proc. nat. Acad. Sci., Wash.*, 1957, **43**, 122

[45]La Cour, L. F. and Pelc, S. C. *Nature, Lond.*, 1958, **182**, 506

[46]Woods, P. S. and Schairer, M. U. *Nature, Lond.*, 1959, **183**, 303

[47]Taylor, J. H. *Exp. Cell Res.*, 1958, **15**, 350 and in *Cell Physiology oj Neoplasia*, p. 547. Austin; Univ. of Texas Press, 1960

[48]Swanson, C. P. *Cytology and Cytogenetics.* London; Macmillan, 1958

[49]McClintock, B. L. *Genetics*, 1944, **29**, 478

[50]Mazia, D. *Proc. nat. Acad. Sci., Wash.*, 1954, **40**, 521; Ambrose, E. J. *Progr. Biophys.*, 1956, **6**, 25

[51]Kauffman, B. P., Gay, H. and McDonald, M. R. *Int. Rev. Cytol.*, 1960. 9, 77

[52]Steffensen, D. M. *Int. Rev. Cytol.*, 1961, **12**, 163

[53]Schwartz, D. *J. Cell. Comp. Physiol.*, 1955, **45**, Suppl. 2, 171

[54]Freese, E. *Cold Spr. Harb. Symp. quant. Biol.*, 1958, **23**, 13

[55]Stadler, L. J. and Uber, F. M. *Genetics*, 1942, **27**, 84; Kirby-Smith, J. S. and Craig, D. L. *Genetics*, 1957, **42**, 176

[56]Shore, V. G. and Pardee, A. B. *Arch. Biochem. Biophys.*, 1956, **62**, 355

[57]Meselson, M. and Stahl, F. W. *Proc. nat. Acad. Sci., Wash.*, 1958, **44**, 671

[58]Levinthal, C. and Crane, H. R. *Proc. nat. Acad. Sci., Wash.*, 1956, **42**, 436

[59]Longuet-Higgins, H. C. and Zimm, B. H. *J. mol. Biol.*, 1960, **2**, 1

[60]Kellenberger, E., Séchaud, J. and Ryter, A. *Virology*, 1959, **8**, 478

[61]Mahler, H. R. and Fraser, D. *Nature, Lond.*, 1961, **189**, 948

[62]Dunn, D. B. and Smith, J. D. *Fourth Int. Cong. Biochem.*, p. 72. 1959

[63]Bendich, A. and Rosenkrantz, H. S. In *Progress in Nucleic Acid Research*, Vol. 1, (Ed. Cohn and Davidson) p. 219, New York; Academic Press, 1962

[64]Taylor, J. H. In *Molecular Genetics Part 1*, (Ed. Taylor) p. 65, New York; Academic Press, 1963

[65]Taylor, J. H. *Genetics*, 1958, **43**, 515

PART 3

THE STRUCTURE OF THE NUCLEIC ACIDS IN RELATION TO THEIR BIOLOGICAL FUNCTION

CHAPTER 8

REPLICATION

Any explanation of the facts of heredity in terms of molecular structure must eventually account for self-duplication, for specificity, and for variation and modification of the genetic material. The helical structure of DNA has certain implications, some or all of which may prove to be important to its biological function: they are the subject of this and the following chapters.

The double-helical structure allows, in principle, the possibility of self-duplication. In the schematic representation of *Figure 8.1*, if the

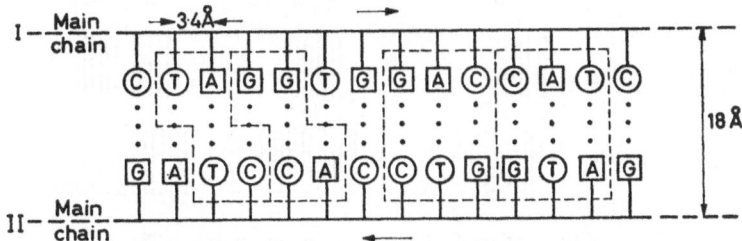

Figure 8.1. Diagram of two possible complementary base sequences in the double-helical deoxyribonucleate molecule. □ Purine base (A = adenine residue, G = guanine residue). ○ Pyrimidine base (T = thymine residue, C = cytosine or 5-methylcytosine residue). ⋮ Hydrogen bond systems (see Formulae IX and X) cross-linking the two helical main chains, represented by continuous straight lines. Arrows show directions of the main chains. Helix diameter 18 Å; the internucleotide distance of 3·4 Å along the axis refers to the para-crystalline B form. Broken lines on the left show an overlapping code; those on the right a non-overlapping triplet code

sequence I is CTAGGTG. . . and II is the complementary sequence GATCCAC. . . , then if I and II can be separated in some way and acquire new partners in two new double helices, these will have the structures I–II' and I'–II, respectively, where the primes indicate new chains. The sequence of nucleotide pairs of the original molecule I–II has, therefore, reproduced itself in its 'progeny', I–II' and I'–II. Since it is not possible to separate laterally two coils entwined as in the DNA structure, there has been much speculation on possible methods for the replication of DNA. As originally conceived[1], the

91

two strands of DNA were thought to separate during duplication and to attract free nucleotides, which were then built up into helices complementary to each of the original strands[1]. This, however, raised the question of how the two intertwined strands of the parent DNA could be separated and at the same time could maintain the helical configuration sufficiently intact to act as the template for a new chain. Although at first it appeared mechanically very unlikely, theoretical calculations show that it is statistically feasible for the DNA to unwind and that the time required is only of the order of a few seconds[2], if the molecular

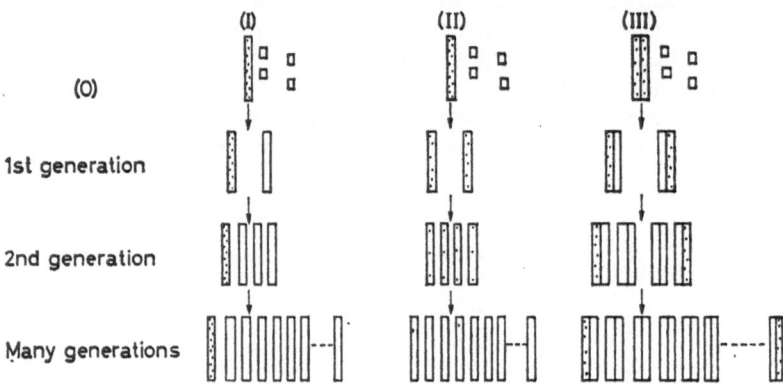

Figure 8.2. Three types of model for the replication process[5]. The dots represent radioactive label and the open squares the non-radioactive sub-units used to build the new structure. A rectangle represents one-half of a replicating particle; in the case of DNA this corresponds to one of the helical chains: (O): original labelled particle. I: template type replication, which leaves the label in one particle. II: dispersive type of replication. III: complementary type of replication

(Reproduced by courtesy of the Editor of *Proceedings of the National Academy of Sciences of the United States of America*)

weight is 6×10^6; correspondingly longer times will be required if the molecular weights are 10^8 or more. These calculations and the demonstration (Plate 3 and page 109) that strand separation can actually be observed meets some of the objections[3] which may be raised against the proposal of Watson and Crick. The work of Kornberg and his colleagues on a DNA synthesizing system which needs intact DNA as a template supports the general correctness of these ideas, and has already been described (Chapter 4, page 38).

The complementary method of replication, which implies two sub-units within the original particle is, however, only one of the possible types of replication which may be envisaged. Other possibilities are depicted schematically in *Figure 8.2*. I is a *template* type of replication (or 'conservative replication'[3]) which leaves the original atoms of the parent particle intact throughout subsequent steps. If the original particle were labelled by radioactive atoms (for example, ^{32}P or tritium), these labelled atoms would continue to remain in the one particle, regardless of the number of replicas formed. II is a *dispersive* type of replication in which the original (labelled) atoms become distributed equally between the daughter particles at each replication, finally leading to a state in which very few labelled atoms would be found in any one particle[4]. III is the *complementary* type of replication suggested by Watson and Crick for the DNA molecule[1]. The labelled atoms exist in two distinct sub-units in the parent particle, separate at the first duplication and each then acquires non-labelled partners. As further duplication occurs, each of the original sub-units remains intact so that, however many such further steps occur, two of the particles present will always contain half the labelled atoms of the parent particle, while the rest of the progeny is completely unlabelled. This type of replication has been called '*semi-conservative*'[3]. The different schemes may apply to the replication of DNA molecules, of larger units containing many DNA molecules (for example, in bacteriophage), or of complete chromosomes—all denoted by 'particle' in the foregoing account.

Much caution is needed in making deductions from the distribution of labelled atoms among the progeny of a labelled particle replicating in an unlabelled environment when genetic recombination can occur[3, 5]. The only experimental result capable of unambiguous interpretation is one in which the integrity of either the entire parental DNA, or of each of its two chains separately, is found to have survived many successive acts of replication and recombination[3]. In such a case it could be inferred that replication proceeded by mechanism I or III, and that recombination occurred without fragmentation of the parental particle. Otherwise it is very difficult to make any deductions, and these limitations must be remembered in assessing such experimental evidence as is available,

The duplication of the labelled DNA of bacteriophage has been studied in recent years, since this system seems to be one of the most favourable for examining together DNA replication, protein synthesis and transfer of genetic information[3, 6, 7]. The distribution of 'parental DNA' (that is, DNA in the original infective phage) is best determined by a method which measures the amount of any original labelling in *individual* progeny phage. Levinthal and Thomas[8] utilized the β-ray emission of ^{32}P incorporated into the parental phage DNA and the technique of autoradiography to count the ^{32}P content of individual T2 phage in the progeny after infection of *E. coli*. These experiments showed that amongst the progeny of ^{32}P-labelled phage, developing in the absence of the label, the phage particles still labelled in the first and second 'generation' each possessed about 20 per cent of the label of its parent. (Note each 'generation' in this context means a multiplication of the original DNA by 30–50 before the appearance of mature phage.) Moreover, experiments on the release by osmotic shock of phage DNA from intact labelled parent phage showed that about 40 per cent of each T2 phage particle could be dispersed by this means and that this 'big piece', of molecular weight about 45 million, was transferred to the first and second generation progeny. Since only these larger particles were detected by the methods of autoradiography, the figure of 20 per cent already mentioned meant that among the first and second generation progeny there were a few particles in which this large piece has approximately half the ^{32}P atoms of the corresponding 'big piece' in the parental phage, in accordance with complementary, semi-conservative replication of type III (*Figure 8.2*).

Since these observations, it has been proved[9] that the DNA in each individual T2 bacteriophage is one entire DNA molecule of molecular weight between 130×10^6 and 160×10^6. Even the 40 per cent 'big piece' is the result of degradation, probably by shearing forces. Nevertheless, the results on the transfer of ^{32}P to intact phage after labelled phage particles have infected single cells are not modified by this discovery, and it may still be concluded that a significant fraction of the label is transferred semi-conservatively in a large fragment. However, any interpretation of the mechanism of recombination which is based on the assumption that the 40 per cent piece was the intact chromosome is no longer sound. Indeed, contrary to earlier interpretations, experiments on recombinant bacteriophages, suitably labelled with ^{13}C and ^{15}N and examined by density-gradient centrifugation (as described in the next paragraph), have shown that the DNA of this bacteriophage is not only contained in a single, semi-conservative replicating chromosome, but also that recombination occurs by chromosome breakage at points along the genetic map proportional to the amounts of DNA in the phage chromosome[10].

More direct support for this type of replication at the molecular level has been obtained by the elegant experiments of Meselson and Stahl[11]. By ultra-centrifugation to equilibrium in a density gradient of caesium chloride they distinguished between DNA molecules which contained ^{15}N or ^{14}N, or equal amounts of both, and to measure the approximate proportion of each type. In a population of *E. coli* uni-

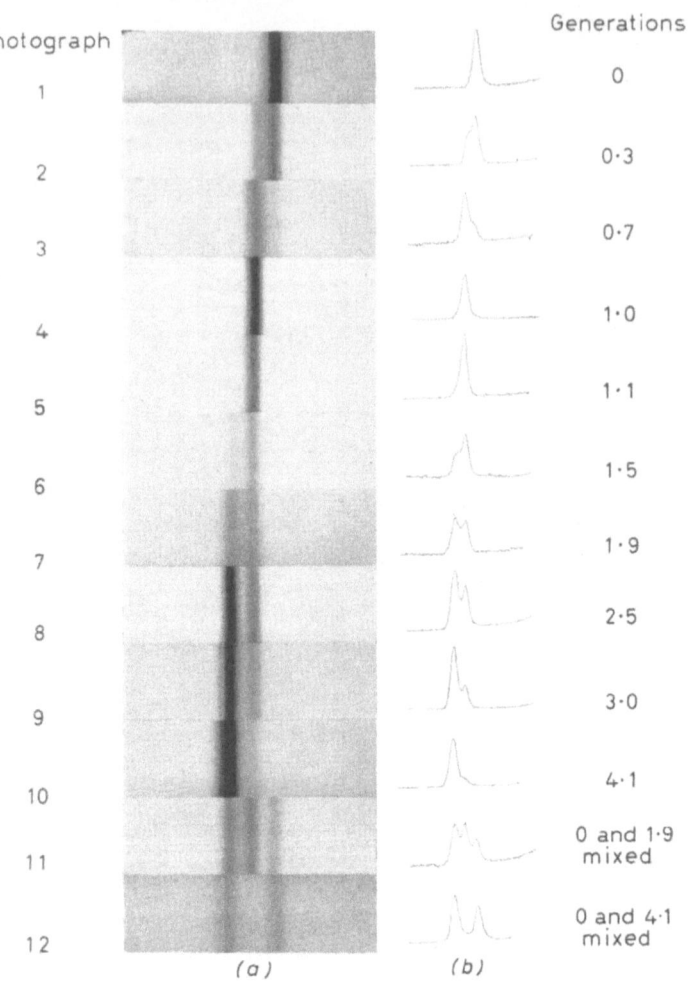

Photograph		Generations
1		0
2		0·3
3		0·7
4		1·0
5		1·1
6		1·5
7		1·9
8		2·5
9		3·0
10		4·1
11		0 and 1·9 mixed
12		0 and 4·1 mixed

(a)   *(b)*

PLATE 3

Ultra-violet absorption photographs[11] showing DNA bands resulting from density–gradient centrifugation of lysates of bacteria sampled at various times after the addition of an excess of ^{14}N substrates to a growing ^{15}N-labelled culture. The density of the CsCl solution increases to the right. Regions of equal density occupy the same horizontal position on each photograph. The time of sampling was measured from the time of the addition of ^{14}N in units of the generation time. b: Microdensitometer tracings of the DNA bands are shown on the right and are directly proportional to the concentration of DNA. The degree of labelling of a species of DNA corresponds to the relative position of its band between the bands of fully labelled and unlabelled DNA shown in the lowermost frame, which served as a density reference. A test of the conclusion that the DNA in the band of intermediate density was just half-labelled was provided by the frame showing the mixture of generations 0 and 1·9. When allowance was made for the relative amounts of DNA in the three peaks, the peak of intermediate density is found to be centered at 50 ± 2 per cent of the distance between the ^{14}N and ^{15}N peaks

(Reproduced by courtesy of the Editor of *Proceedings of the National Academy of Sciences of the United States of America*)

To face p. 95

formly labelled with ^{15}N, the distribution of ^{15}N amongst molecules of bacterial DNA could be followed while the cells were growing exponentially in ^{14}N medium (Plate 3). Density gradient analysis of DNA in lysates of the initial bacterial population revealed only one band corresponding to ^{15}N-labelled DNA (Plate 3/1). During growth in ^{14}N-medium, hydrid half-labelled particles (containing ^{15}N and ^{14}N in approximately equal quantities) accumulated until at the end of one generation time such molecules were the only ones present (Plate 3/2–5). Subsequently only hybrid and completely unlabelled DNA were found, the quantities of the two types being the same after two generation times (Plate 3/7). From this it can be concluded that the nitrogen of the DNA containing particles is divided equally between two sub-units which remain intact for many generations and that during replication each daughter particle receives one parental sub-unit, as in *Figure 8.2*, III, where the dots now represent the ^{15}N-labelling. These results therefore rule out the dispersive type of replication for the particles containing DNA and are in accord with the complementary semi-conservative type.

Exactly parallel observations have been made[12] on Chlamydomonas and semi-conservative replication of the DNA of mammalian cells (epithelial cells[13] and human HeLa, cervical carcinoma cells[14]) has also been proved by similar density gradient sedimentation of the DNA after it has incorporated 5-bromodeoxyuridine, which has the advantage of causing a larger density difference. However, these observations did not prove conclusively that the replicating unit was indeed the double-helical molecule and that the progeny were hybrid double helices containing one new and one old polynucleotide strand.

Cavalieri and Rosenberg[15] have argued that their observations on the effects of heat and enzymatic degradation on the viscosity, molecular weight and behaviour in caesium chloride gradients of a variety of DNA indicate that DNA from proliferating sources contains two separable units and DNA from non-proliferating sources only one such unit. The important point is that these conserved units are considered to be undenatured double-helical DNA, so that the DNA from proliferating sources contains altogether 4 polynucleotide strands, as a pair of double helices. They propose that replication of DNA is an alternation between the four-stranded forms and two-stranded forms, the latter being the conserved unit. Replication of the parental four-stranded system would then be semi-conservative but that of the double helix of DNA would be conservative.

These suggestions differ from the original scheme of Watson and

Crick[1] who thought of the single polynucleotide strand as the conserved unit and, indeed, the supposed biological role of specific base pairing depends on this. This molecular mechanism of transference of genetic information from one 'generation' of DNA to the next is obscure in the proposal of Cavalieri and Rosenberg. Their suggestion has provoked further studies and some of the reasons for thinking that the DNA from all types of sources is two- and not four-stranded have been mentioned earlier (Chapter 4, page 33). It seems likely that double helices of DNA are laterally associated in the chromosome, and it may well be that these workers have succeeded in isolating DNA from chromosomes before it has separated into separate double-helical molecules.

The important question is: What is the molecular structure which actually replicates? Cairns[16] pointed out that a decision could be reached if some DNA were found which (a) becomes hybrid on replication and (b) can be isolated intact in its entirety and shown to be two-stranded in this state. Bacterial DNA was unsuitable because, although shown to be hybrid on replication[11], as already described, there was at that time no unambiguous estimate of its native length-to-mass ratio, from which the number of strands per molecule could be inferred. T2 bacteriophage was also unsuitable because, although known to be two-stranded[17], only a small part of the total was known to be hybrid and these could be four-stranded without affecting very much the average mass-to-length ratio. However λ-bacteriophage DNA was known to be transferred[10] to progeny particles which were hybrid, that is, half parental and half new, and all the DNA exists in this bacteriophage as a single molecule.

As described earlier (Chapter 4, page 35), Cairns measured by autoradiography the length of λ-bacteriophage molecules which had replicated. This value (of 23 μ) combined with the reported molecular weight of 46×10^6 was the mass-to-length ratio for the double helix. The only weakness remaining in this argument is that the value assumed for the molecular weight was inevitably obtained indirectly from sedimentation studies (Chapter 4, page 35). This is not a very serious objection and the two-strandedness of the replicated λ-DNA was, in any case, confirmed by the density per unit length along the autoradiograph of grains from the tritiated thymine. The hybrid λ-DNA formed on replication is therefore two-stranded and replication must therefore have involved separation of the two original polynucleotide strands, as Watson and Crick proposed. The significance for replication of the observation of Cavalieri and Rosenberg is thus uncertain, although their observations are very relevant to the problems of the organization of DNA in the chromosome and of the discrepancies between molecular weights of different DNA preparations.

Doubts have sometimes been expressed concerning the Watson and Crick replication mechanism since it was thought to be unlikely that the complementary base-pairing would be specific enough to avoid errors. Nevertheless when a copolymer of adenine and thymine is used as a primer for DNA polymerase (Chapter 4, page 38) in the presence of ^{32}P-labelled deoxyguanine triphosphate and the other, unlabelled, deoxyriboside triphosphates, less than one guanine in the presence of 28,000 thymine was incorporated into the new poly-deoxyribonucleotide[18]. The result was even more unequivocal when

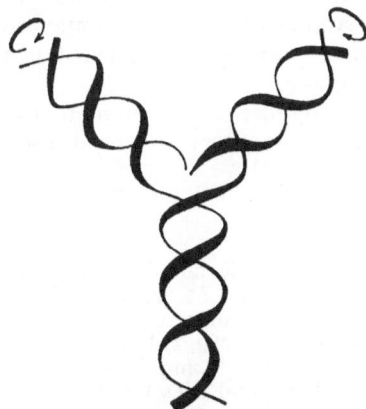

Figure 8.3. Simultaneous unwinding and replication of DNA[3]
(Reproduced by courtesy of Johns Hopkins Press, Baltimore)

the primer was a polymer of adenine and bromouracil. Guanine cannot pair with either adenine or thymine (= bromouracil) and so provides a sensitive test of the accuracy of the specific pairing during replication.

There are many unsolved questions concerning the replication mechanism, even if it is accepted that it occurs by duplication of the individual strands of the double helix. It is not, for example, clear whether or not the two strands separate fully before duplication, each strand then being 'processed' by an enzyme adding on a new complementary chain; or whether the unravelling of the parent DNA occurs simultaneously with the formation of the two daughter helices as depicted[3] in the scheme of *Figure 8.3*, the requisite free energy coming from the breakdown of nucleoside triphosphates into monophosphates. Strand separation[19] can occur *in vitro* under the appropriate conditions of low concentration and high ionic strength but the relation between such studies and the conditions prevailing *in vivo* is not clear.

The availability of tritium-labelled thymidine has led to studies of replication in chromosomes which have been more fully described in Chapter 7. However, the fundamental difficulties in drawing conclusions about DNA replication from studies of larger units such as the chromosome have been summarized by Delbruck and Stent in the following warning[3] which needs to be borne in mind.

'An observed *unequal* distribution of parental atoms may be due to an unequal role in the duplication process of various fractions of the total DNA contained in the self-duplicating structure, while an observed *equal* distribution may only reflect the randomizing effect of some post-replication event like fragmentation by genetic recombination.'

Reference should be made to Chapter 7 for a fuller discussion of the semi-conservative replication of chromosomes.

REFERENCES

[1] Watson, J. D. and Crick, F. H. C. *Nature, Lond.*, 1953, **171**, 737

[2] Wolkenstein, M. V., *Proc. Vth Intern. Congr. Biochem.*, 1961, p. 100, Moscow, Symposium No. 1; Longuet-Higgins, H. C. and Zimm, B. H., *J. mol. Biol.*, 1960, **2**, 1

[3] Delbruck, M. and Stent, G. S. *The Chemical Basis of Heredity*, (Ed. McElroy and Glass), p. 699. Baltimore; Johns Hopkins Press, 1957

[4] Delbruck, M. *Proc. nat. Acad. Sci., Wash.*, 1954, **40**, 783

[5] Levinthal, C. *Proc. nat. Acad. Sci., Wash.*, 1956, **42**, 394

[6] Burton, K. *Biochem. Soc. Symp.*, 1957, **14**, 60

[7] Hershey, A. D., Garen, A., Fraser, D. K. and Hudis, J. D. *Carnegie Inst. Year Book*, No. 53, p. 210. 1954

[8] Levinthal, C. and Thomas, C. A. *The Chemical Basis of Heredity*, (Ed. McElroy and Glass), p. 737. Baltimore; Johns Hopkins Press, 1957

[9] Davison, P. F., Freifelder, D., Hede, R. and Levinthal, C. *Proc. nat. Acad. Sci., Wash.*, 1961, **47**, 1123; Rubenstein, I., Thomas, C. A. and Hershey, A. D. *Proc. nat. Acad. Sci., Wash.*, 1961, **47**, 1113; Cairns, J. *J. mol. Biol.*, 1961, **3**, 756

[10] Meselson, M. and Weigle, J. J. *Proc. nat. Acad. Sci., Wash.*, 1961, **47**, 857

[11] Meselson, M. and Stahl, F. W. *Proc. nat. Acad. Sci., Wash.*, 1958, **44**, 671

[12] Sueoka, N. *Proc. nat. Acad. Sci., Wash.*, 1960, **46**, 83

[13] Chun, E. H. L. and Littlefield, J. W. *J. mol. Biol.*, 1961, **3**, 668

[14] Simon, E. H. *J. mol. Biol.*, 1961, **3**, 101

[15] Cavalieri, L. F. and Rosenberg, B. H. *Biophys. J.*, 1961, **1**, 301, 317, 323, 337

[16] Cairns, J. *Nature, Lond.*, 1962, **194**, 1274

REFERENCES

[17] Cairns, J. *J. mol. Biol.*, 1961, **3**, 756
[18] Trautner, T. A., Swartz, M. N. and Kornberg, A. *Proc. nat. Acad. Sci., Wash.*, 1962, **48**, 449
[19] Marmur, J., Rownd, R. and Schildkraut, C. L. *Progress in Nucleic Acid Research*, 1963, **1**, 232

CHAPTER 9

THE MODIFICATION OF NUCLEIC ACID STRUCTURE

There is a particularly close and immediate relation between the biological controlling function of the nucleic acids and their precise molecular structure. This activity is, in the broadest sense, genetic and it is natural to expect it to be altered when the structure of the nucleic acid is modified. Structural changes may alter the ability of DNA to replicate or of the DNA or RNA to code correctly for specific proteins and this is eventually expressed in altered phenotypic forms, the changes varying from the trivial to the lethal. It has been the object of much recent work to relate the biological effects of a variety of agents to the parallel structural changes which they induce in the nucleic acids. In this way it is hoped to determine the molecular basis of hereditary changes.

The changes which may occur in the structure of DNA are represented schematically in *Figure 9.1*, where they are classified as* (α) breaks in the main chain phospho–ester linkages, (β) breaks in other covalent bonds of the nucleotide units, (γ) rupture of the interchain hydrogen bonds, (δ) formation of covalent cross-links. The possible changes in RNA are the same, since the figure may be regarded as representing a short helical section of RNA and includes the changes in single chains.

RUPTURE OF THE INTERNUCLEOTIDE PHOSPHO–ESTER LINKAGES†

It would be expected that the double helix would only separate into two parts if the breaks occurred at opposite, or nearly opposite, positions in the two chains ($\alpha 1$, $\alpha 2$). Such a change, with a decrease in molecular weight, is called *degradation*. If the breaks in the opposite chains (for example, $\alpha 1$, $\alpha 4$) are separated by a sufficient number of hydrogen-bonded pairs, the molecule remains intact, although its shape may change. Breaks in the same chain (for example, $\alpha 1$, $\alpha 3$)

---

* The notation α, β γ, δ has no relation to the description of various ionizing radiations as α, β, etc.
† $\alpha 1$–$\alpha 5$, *Figure 9.1*.

need not cause detachment of the intervening units and would be revealed only by any subsequent rupture of the hydrogen bonds.

It is possible to find a *formal* similarity between the results of inter-nucleotide bond rupture in DNA and chromosomal changes, when these are both caused by the same agent, such as ionizing radiation. Thus, if breaks occurred at α1, α2, α4 and α5 (*Figure 9.1*), the section α1–α2 and α4–α5 would be deleted and this might be followed by the joining of positions α1 to α5 and α2 to α4. This would then be formally similar to deletion within a chromosome and would be expected to have genetic consequences since nucleotide sequence is the basis of gene action and of protein specificity. Similarly, the double break α1–α2 would result in two fragments which, if they linked up with

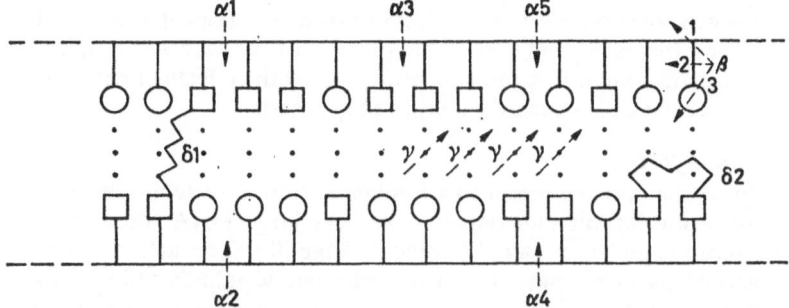

Figure 9.1. Possible modifications of the DNA molecule (depicted schematically as in Figure 8.1). --→ Possible breaks. α, breaks in the main chain phospho–ester linkages. β, breaks in other covalent bonds of the nucleotide units. γ, rupture of the inter-chain hydrogen bonds. δ, formation of covalent cross-links

others, would give an effect analogous to chromosomal interchange. Furthermore, if the breaks α3 and α5 occurred in conjunction with hydrogen-bond rupture (γ, *Figure 9.1*) the segment α3–α5 would be lost and the replication of this section would be impaired: this could be part of the molecular basis for mutations.

At present, these can constitute only formal analogies, because of the disparity of scale between chromosomes and DNA molecules (Chapter 7). If, however, the structure of the 200 Å microfibril of the chromosome proves to consist of parallel bundles of identical duplicated nucleoprotein molecules, then the above may no longer be analogies but the molecular events which become magnified into optically visible changes at the cytological level. (For a discussion of this problem *see* Freese[1].)

The agents which cleave internucleotide linkages are as follows.

Ionizing radiation[2-5]

Ionizing radiation whether acting indirectly through free radicals (H, OH· and O_2^-), produced in dilute aqueous solutions of the nucleic acid, or directly on nucleic acid as a gel or solid. X-rays, γ-rays and high-energy electrons produce only single internucleotide breaks in DNA and, in order to cause degradation, approximately opposite breaks have been shown to be necessary[3]. Densely ionizing particles, such as α-rays, produce ionizations very close to each other and the passage of only one particle can cause a double break.

Sonic and ultrasonic waves[6]

Scission of both chains of DNA at opposite points with relatively few single breaks occurs with, at first, little denaturation[6a,7]. The effect of these waves is probably, in the first instance, mechanical, through the shearing forces engendered in the passage of the waves through the solution, although free radicals produced by the effects of cavitation on water have also been thought to play a part.

Shearing forces[8]

Even the stresses present in forcing a solution of DNA through a hypodermic needle can degrade the very large DNA molecules of bacteriophage virus (*see* Chapter 4, page 35). The action may be regarded as mechanical and the molecular weight is reduced to a lower limiting level dependent on the shearing forces. The effects seem to be entirely degradative through the production of double breaks, although this aspect has not been studied in detail.

Specific enzymes[9] (*deoxyribonuclease for DNA and ribonuclease for RNA*)

The phospho–ester bonds of the nucleic acids are hydrolysed with liberation of secondary phosphoryl groups and a decrease in molecular weight. Only single bonds are attacked at a time, and again two nearly opposite breaks are required for a decrease in molecular weight[10]. Study of the time course of this degradation showed that the single breaks do not have to be exactly opposite for there to be a decrease in molecular weight, separation to the extent of two nucleotides still allowing the two pieces to come apart[10].

Alkylating agents

Alkylating agents such as sulphur mustard, $S(CH_2CH_2Cl)_2$, the nitrogen mustards, $RN(CH_2CH_2Cl)_2$ (R = alkyl radical), and many related compounds attack the base of DNA rings (*see* page 105) and this subsequently leads to fission of the sugar-phosphate backbone[11].

Heating

The primary effect of heat on aqueous solutions of DNA is denaturation, but degradation can occur on heating at high temperatures (> 90° C) for long periods (> 60 min)[12], presumably through hydrolysis of the phospho–ester bonds with addition of the elements of water.

Strong acid and alkali

The primary effect of treatment of DNA with dilute acid and alkali is denaturation. Under more drastic conditions acid releases purines and internucleotide bonds are broken; ultimately free pyrimidines, sugars and phosphoric acid are obtained[13]. Both DNA and RNA can be hydrolysed by alkali, RNA being notably more susceptible.

Succinyl peroxide

Succinyl peroxide, $HO_2C.(CH_2)_2.CO_3H$, causes scission of phospho–ester bonds, but its primary reaction is with the base rings[14].

Since most of the above agents produce other changes in DNA in addition to degradation, it is not often possible to determine a direct relation between degradation and biological activity. Thus ionizing radiation and alkylating agents, both of which cause chromosome breakage and induce mutation, have other structural effects besides degradation (see below). Degradation is not the most important feature of the effect of heat, of extremes of pH, and of succinyl peroxide. However, degradation, without denaturation, or nucleotide breakdown, appears to be the sole result of sonic and ultrasonic irradiation, at least in its earliest stages. Eventually, as with all degradative processes, denaturation must occur as the molecule breaks down into nucleotides. This behaviour has enabled a study to be made of the relation of molecular size of transforming principle DNA to its activity.

DNA which had transforming activity with respect to resistance to antibiotics was isolated from *Diplococcus pneumoniae*, and was degraded to various extents by sonic waves[15]. The logarithm of this activity decreased linearly with the fraction of bonds broken, which occurred at opposite points along the main chains without any disturbance of the remainder of the paired strands. Analysis of the degradation curves by target theory led to a value of one million for the critical molecular weight of DNA for transmission of resistance to each of the three antibiotics. This value may have represented either the minimum size required for effective attachment to a bacterial site or a combination of this and the minimum size required for genetic incorporation following penetration into the cell. In either case this amount of DNA was far greater than the minimum length of the molecule altered in a mutation that gives rise to a genetically distinguishable strain[16].

Occasional internucleotide breaks may be part of the normal DNA structure in the dynamic system of the living cell, although the frequency of these is probably less than the one in 50 which was at one time considered possible. The flexibility which such occasional breaks would introduce seems to be demanded in certain systems, such as the heads of bacteriophage, in which the length along the axis of the isolated DNA molecules is very much greater than the largest dimension of the biological structure from which they are derived[17].

RUPTURE OF COVALENT BONDS WITHIN THE NUCLEOTIDES*

Agents which rupture the sugar or base moieties of the nucleotides would be expected to impair the complementary matching of the nucleotide chains and thereby the duplicating and genetic function of the DNA. Rupture of the heterocyclic base rings naturally leads to eventual breakdown of hydrogen bonding and labilization of the internucleotide linkages is expected to be a consequence of the opening up of the sugar rings of the nucleotides. There are many agents, physical and chemical, which attack the nucleotides and only a selection can be cited.

Ionizing radiation and ultra-violet light

High doses cause the breakdown of nucleic acids with release of ammonia, inorganic phosphate, labile phosphate esters and other degradation products. The reactions occurring include deamination and dehydroxylation of the bases, partial destruction of the base rings, oxidation at the C_4 position of the sugar, and rupture of the base–sugar linkage[4, 5, 18]. These effects were once thought to be of biological significance, but this is improbable because the dosages employed are several orders of magnitude greater than those producing genetic

* $\beta1$, $\beta2$, $\beta3$, *Figure 9.1.*

mutations and also much higher than those causing denaturation or even degradation of the polynucleotide chain. The most important initial effect of ultra-violet light on DNA now appears to be a reaction in which one thymine ring is joined to another by two C—C bonds[19, 20].

Alkylating agents

These compounds attack first the N_7 atoms of the guanine rings and then, less readily, the N_1 and N_3 of adenine and N_1 of cytosine; the alkylated base subsequently splits off by fission of its linkage to the C_1 of the sugar and this is followed by rupture of the sugar–phosphate backbone[11]. Mechanisms involving the initial esterification of the primary phosphate groups have also been postulated but recent work makes these appear less probable.

Nitrous acid

Nitrous acid, HNO_2, reacts with the free amino groups on adenine, guanine and cytosine when they form part of DNA or RNA, and replaces them by an —OH group[20, 22a]. The changes are A→Hypoxanthine $\overset{\text{in}}{\rightarrow}$ G; G→Xanthine; C→U. The relative speeds of these changes differ for RNA and DNA, and the rates $\overset{\text{cell}}{\text{for DNA}}$ depend on whether or not it is denatured and on the pH. There is no decrease in molecular weight as a result of the action of nitrous acid, so it provides a useful way of selectively breaking certain hydrogen bonds without degradation, although there is some loss of purines in the acid solutions inevitably employed (pH 4). Hypoxanthine can pair with cytosine in a double-helical structure as in (poly I + poly C) (Chapter 5, page 58). Although structurally feasible, xanthine appears not to pair with cytosine, since synthetic polyribonucleotides in which guanine has been deaminated by nitrous acid to xanthine, lose their ability to cause amino acid incorporation[22b]. Thus, xanthine cannot replace guanine in the genetic code and, presumably, in base-pairing.

Hydroxylamine

Hydroxylamine, H_2NOH, causes alterations only in the pyrimidines[21]. It opens the ring of uracil in RNA to form a cyclic isoxazolone, leaving a urea residue on the base, and it adds on to cytosine (at position 4 or 6) which is then rapidly deaminated. Hydroxylamine reacts with the pyrimidine 5-hydroxymethyl cytosine in the DNA of T-even phages (instead of cytosine) but not with thymine.

Succinyl peroxide

Both purine and pyrimidine rings are oxidized by succinyl peroxide[14] giving a variety of products, with little rupture of internucleotide or of hydrogen bonds. The DNA is thus modified only with respect to its bases.

The changes in the covalent bonds of nucleic acid which can be induced are so various that no generalizations can be made and each agent must be considered separately.

Ionizing radiation or ultra-violet light have long been known to have a mutagenic effect and to cause chromosome aberrations. The similarity between the action spectrum of ultra-violet light in producing mutation and the absorption spectrum of nucleic acids is similarly well known. Evidence that it is the nucleic acids which are the targets of ultra-violet light in the cell has come from studies of the inactivating effect of the irradiation of transforming principle DNA. The inactivation curves are complex[14, 22, 23], though not multi-hit; that is they are best interpreted as the summed effect of single-hit inactivations of markers of different initial sensitivity. (By a 'marker' is meant transformation with respect to a particular property of the bacteria, for example, resistance to streptomycin). Why markers differ in sensitivity is not clear; it may be connected with their position along the DNA or because more sensitive markers have two operative regions in the DNA and thus a greater chance of inactivation[22].

In contrast, the sensitivity of different markers in transforming principle to the inactivating action of nitrous acid varies very little from one marker to another. The mutagenic action and inactivation of transforming principle[21], the inactivation of viral RNA[22], and deamination by nitrous acid all follow first order (single-hit) kinetics. It was concluded that the deamination of single bases in the DNA or RNA is responsible for inactivation or mutation. Deamination of A, C and G cause inactivation and deamination of A ($\rightarrow$ G, ultimately) and of C ($\rightarrow$ U) can lead to mutations. The target size for inactivation and mutation of *D. pneumoniae* transforming principle were calculated[22] to be approximately 125–200, and about 75–135 base pairs, respectively. However, the target size[21] for the inactivation of TMV RNA by nitrous acid was 3000 nucleotides which corresponds to about half of the molecule, so that the infective unit of this viral RNA cannot be a small sub-unit of the molecule.

Hydroxylamine causes both inactivation and mutation of TMV RNA at alkaline and acid pH, when the reactions with uracil and cytosine are respectively maximal, so the reaction with either pyrimidine base can cause both effects. It induces mutations in T2 and T4 phage according to single-hit kinetics, with increasing efficiency at acid pH, when the reaction with 5-hydroxymethyl cytosine is greatest[21]. Mutation in these bacteriophage is apparently caused by the instantaneous reaction of hydroxylamine with single 5-hydroxymethyl cytosine moieties. Inactivation of phage DNA by hydroxylamine is more complex since it also damages the absorption mechanism in the phage tail.

The ratio of mutations induced to lethal hits is very much greater with hydroxylamine than it is with nitrous acid. Since guanine is not altered by the former but is by the latter it has been suggested[21] that this is the cause of the high lethality of nitrous acid. The deamination product of guanine is xanthine and production of this therefore appears to block the ability of DNA to replicate[24]. This is supported by the observation that guanine cannot be replaced by xanthine in the DNA-polymerizing system of Kornberg (Chapter 4, page 38), although it can be replaced by hypoxanthine (the deamination product of adenine).

Succinyl peroxide inactivates[14] markers of *H. influenzae* transforming principle according to single-hit kinetics. Again this suggests that a single lesion, probably modification of a single base, within a defined region of the DNA, inhibits the transfer of genetic information corresponding to a particular marker. The relative sensitivity to succinyl peroxide of the three markers examined[14] was in the same order as had been observed with inactivation by ultra-violet light, deoxyribonuclease and ionizing radiation. A reasonable interpretation of this is that the sensitivity of the markers is related to the extent of the region of DNA which controls this property.

RUPTURE OF THE COMPLEMENTARY HYDROGEN BONDS OF THE DOUBLE HELIX*

This process, considered apart from other changes is called *denaturation*; it leads to loss of the double-helical configuration, with consequent changes in the viscosity, optical rotation, ultra-violet absorption, titration curves and other physical properties[2,3,6] (*see* Chapter 4, page 30ff). The phenomenon is a co-operative one involving the simultaneous rupture of a group of hydrogen bonds, of the order of 10–20, judging from the activation energy for thermal denaturation[25]. There has been considerable discussion[2] about the extent to

* *γ, Figure 9.1.*

which the process is reversible. Under most circumstances it is not so but there are special conditions of thermal denaturation and subsequent cooling which allow the re-formation of double-helical structures. This process is termed 're-naturation'. Denaturing agents include the following.

Heat

When solutions of DNA are heated there is a critical temperature range over which the DNA is denatured. The temperature corresponding to half-conversion of the helical to random form is called the denaturation temperature or, more colloquially, the 'melting temperature'. This temperature decreases with decreasing GC content of the DNA, with decrease in electrolyte concentration and when urea is present. The time process of denaturation is first order kinetically[25] and at any one temperature proceeds to a level determined by the heterogeneity of composition of the DNA. If heating is prolonged, degradation ensues. Under normal conditions of concentration of DNA and of electrolytes and with rapid cooling, the strands do not separate on heating and there is no change in molecular weight. Thus, by appropriate choice of conditions, heating can be made to effect the transition from double helix to a disorganized form, with rupture only of the complementary hydrogen bonds.

Acid and alkali

Exposure of DNA to solutions of pH 2·5 or 11 causes denaturation through the ionization of the groups on the bases which are involved in the hydrogen bonds[6, 26]. The phenomenon is usually irreversible at 25° C, but it becomes reversible if the ionization is effected at temperatures near to 0° C. Changes in many physical properties[2, 3, 6, 27] confirm that the helical form is lost without any degradation at pH 2·5 or 11 at 25° C, if the time the solutions are held at these pH values is not too long (not more than about 30 min).

Low ionic strength[6, 27]

If the concentration of sodium chloride in solutions of DNA is lowered to below about 10^{-3}M, irreversible denaturation occurs at room temperature, probably because the increased repulsion between the phosphate charges on the DNA renders the structure unstable. Dilution of pure DNA solutions has the same effect.

Urea and guanidinium chloride

Urea and guanidinium chloride[6] denature DNA by forming hydrogen bonds directly with groups on the base rings, thereby replacing the inter-base hydrogen bonding required to maintain the double-helical structure.

Ionizing radiation and ultra-violet light

One of the first detectable structural effects of γ-rays on DNA in aqueous solution (indirect action) is the rupture of hydrogen bonds[28]. The evidence suggests that this effect is not random, the hydrogen bonds linking adenine and thymine being broken more readily than those linking guanine and cytosine. The process is far more efficient than rupture of internucleotide linkages, about 10–20 base pairs being severed through rupture of hydrogen bonds for each internucleotide break in the main chain. Partial denaturation of DNA is also caused by ultra-violet light[29].

Changes of type β

Denaturation is only a secondary effect of the action of enzymes, nitrous acid, hydroxylamine, alkylating agents, and of sonic and ultra-sonic irradiation and is, in these cases, simply consequential upon breakdown of the chemical structure through processes of type β described above.

It would be expected that the opening up of sequences of hydrogen bonds with loss of the double-helical structure would impair the ability of DNA to replicate over the region denatured. This region should then have a reduced viability or might become the locus for a permanent mutation. The loss of transforming activity of *H. influenzae* when subjected to heat, changes of pH, or of ionic strength is paralleled by its loss in viscosity, caused by collapse of the helical structure on denaturation[30]. The classical observation of this kind is shown in *Figure 9.2*. This relation between helical structure and transforming activity has been confirmed and elaborated in many other cases, for example with the transforming principle DNA of *D. pneumoniae*. The transforming DNA labelled with respect to different markers (resistance to antibiotics) displayed different susceptibilities to heat denaturation, which means that the specific marker regions of the DNA have different average base compositions, because of the relation, already mentioned, between denaturation temperature and GC content. When DNA of bacterial origin is heated under carefully specified conditions the two strands separate, but on slow cooling to room temperature they can partly recombine to form a double helix again[31]. This 're-natured' DNA can also be formed by the union of

109

strands obtained by heating mixed solutions of DNA from genetically related but distinct bacteria, and by heating mixed solutions of 'marked' DNA and of 'wild-type' DNA. Thus, there is now the interesting possibility of forming 'heterozygous' DNA by renaturation, that is, DNA molecules with different genetic markers or chemical modifications in the two strands. It has already been shown that hybrid molecules in which one strand contains a mutant marker

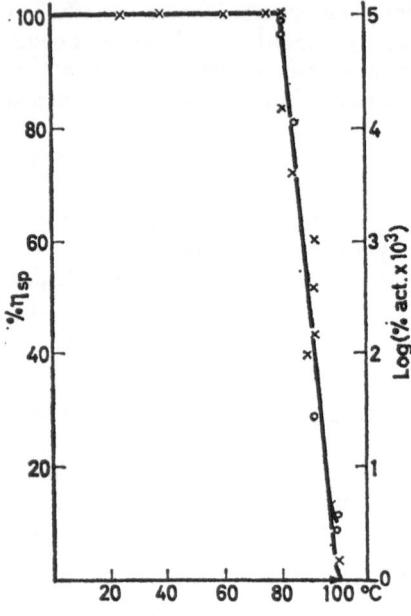

Figure 9.2. The parallel changes in viscosity (o), as percentage of original specific viscosity, and in transforming activity (x), as log (% of maximum activity × 10³), of transforming principle DNA from Haemophilus influenzae as a function of temperature of previous heating for 1 hour[29]. Transforming activity was determined with respect to polyribophosphate production and to streptomycin resistance

(Reproduced by courtesy of the Editor of *Journal of Experimental Medicine*)

and the other is derived from wild-type DNA are active in carrying out transformations with respect to the mutant property[32]. This strongly supports the view that the essential genetic information in DNA is carried independently by each strand.

At the cytological level it is interesting to note that those agents which loosen the *visible* spiralling of chromosomes[33], namely, heat shock, nitric acid vapour and ammonia gas[34] are precisely those which would be expected to disorganize the *molecular* helices of the DNA molecule through rupture of hydrogen bonds by heat denaturation, decrease and increase of pH, respectively. The double helices of DNA can now reasonably be identified with the 'molecular spirals' long postulated on mechanical grounds as the basis of the visible chromosome spirals[34].

CROSS-LINKING THROUGH COVALENT BONDS*

Some agents introduce groups or remove hydrogen atoms so that one nucleotide is joined to another by new covalent bonds which thereby link two polynucleotide chains or different parts of the same chain. If the two chains which are linked are in the same double helix ($\delta 1$, *Figure 9.1*) then the separation of the two strands is made more difficult. Alternatively, the two strands which are cross-linked may belong to different double helices, in which case there is a tendency for gels or insoluble fibres to be formed. Cross-linking effects have been observed with the following.

Ionizing radiation

The direct action of ionizing radiations on solid DNA at large dosages causes cross-linking of DNA molecules[35], as well as degradation and denaturation. This has been observed when the nuclei of herring sperm are irradiated by α-rays, β-rays or x-rays and appears to involve covalent cross-linking between different DNA molecules (double helices), but not between protein. Such cross-linking would explain the earlier observations of an increase in molecular weight at higher dosages[36] and the tendency of solid DNA to form gels in solution after extensive irradiation[37].

Ultra-violet light

Irradiation of DNA with ultra-violet light causes cross-linking between thymine rings and these probably constitute the intramolecular cross-links which it seems to induce in various bacterial DNA. The presence of such intramolecular, interchain, cross-links was inferred when it was found that heat or formamide treatment, which breaks hydrogen bonds, could not induce the strand separation which could be obtained under the same experimental conditions with DNA not previously irradiated with ultra-violet light[29]. (Strand separation can be followed by means of sedimentation to equilibrium in a gradient of caesium chloride of DNA labelled in one chain.) The question of intrachain links ($\delta 2$, *Figure 9.1*) has not been settled.

Alkylating agents

The most effective alkylating agents are difunctional, those with two reactive alkylating groups per molecule. Each group can alkylate a

* $\delta 1$ and $\delta 2$, *Figure 9.1*.

guanine ring at the N_7 position so that one molecule of alkylating agent joins two guanine residues[11]. The evidence so far provisionally favours cross-linking which is interchain and intramolecular ($\delta 1$, *Figure 9.1*). This is possible since a chain of five atoms (—C—C—N—C—C—) is just long enough to join the N_7 atoms of the guanine rings, provided these were members of adjacent base pairs, so the sequence

```
  |       |
  G ···   C
  |  \    |
  C ···   G
  |       |
```

is needed for such cross-links to be formed. (The arrowed line is the cross-link.) There is still a possibility that the bifunctional alkylating agents join the N_7 atoms of guanine rings on nucleotides adjacent in the same chain ($\delta 2$, *Figure 9.1*), and this requires the sequence

```
  |       |
  C ···   G
  |       | )
  C ···   G
  |       |
```

Nitrous acid

There is evidence that nitrous acid introduces into DNA covalent cross-links between chains of the double helix (type $\delta 1$). Such cross-links render denaturation reversible by holding the disordered chains together[38].

The biological effects of the cross-linking agents have been discussed in connexion with the other changes they produce in the structure of nucleic acids and it is difficult, especially in the case of ultra-violet light, to know how much of their biological effects are to be attributed to cross-linking alone. (Cross-linking is unimportant at low doses of ionizing radiation.) Clearly any interchain cross-links between bases would prevent the bases from pairing properly during replication and any interchain intramolecular cross-links would prevent strand separation and thereby replication.

The difunctional alkylating agents afford an opportunity of determining the biological effects of cross-linking by comparing them with their monofunctional analogues. The difunctional compounds are certainly more effective cytotoxically; they yield a higher proportion of chromosome breaks, but a lower proportion of single chromatid mutants. This can be understood if the difunctional agents attack *both* chains of the DNA and if changes in DNA lead to mutagenic effects and scission of DNA to chromosome breaks. The effects of the monofunctional and difunctional alkylating agents are illustrated in *Figure 9.3*. Lawley[11] has listed the following possibilities:

(1) Alkylation of a base ring alters its tautomeric structure so as to alter or prevent its pairing with another base.

(2) The loss of the alkylated base moiety by hydrolysis results in deletion of the base-pair in which it occurred.

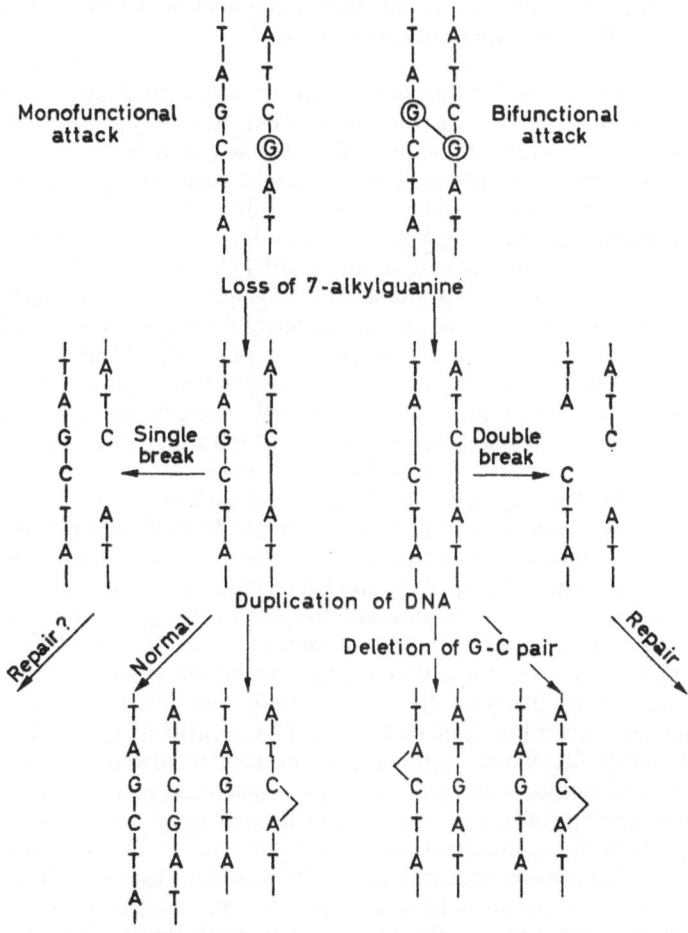

Figure 9.3. Diagrammatic representation of the possible effects of loss of alkylated guanine moieties on the reduplication of DNA[11]

(Reproduced by courtesy of the Editor of *Journal de chimie physique*)

(3) The lost base is replaced wrongly by a repair mechanism.

(4) The loss of the base and subsequent chain fission is repaired wrongly.

Not many quantitive correlations between biological activity and the chemical effects of alkylating agents on the nucleic acids are available. Alkylation of TMV to the extent of only a few moles per particle weight of RNA is sufficient to inactivate it[39]. The difunctional compounds are much more effective in inactivating T2 bacteriophage than are the monofunctional compounds[40].

A survey such as the foregoing of various agents and their manifold effects on nucleic acid structure soon reveals that very few can be said to have one particular unique effect which can then be studied in isolation from other possibilities. In recent years, the growth in our understanding of the structural changes has been matched by the availability of an increasing number of biological tests of activity. The most studied are: the infectivity of viral RNA (especially TMV); the activity of transforming principle DNA; genetic changes in bacteriophage DNA; and effects on the amino acid sequence of proteins dependent on DNA or RNA (*see* Chapter 12). Most of these biological effects include a number of stages (for example, penetration of transforming principle into the cell or attachment of bacteriophage to it), and it is not always possible to tell which of them is being affected by the modification in the nucleic acids.

However, the gap between genetic event and molecular change is continually being narrowed. Notably, it has been shown possible to obtain recombination between very closely linked sites on the genetical chromosome of T4 *E. coli* bacteriophage[16]. This has enabled an attempt to be made to correlate the size of the 'gene' (in terms of recombination units) with the physical size of the DNA[40]. This bacteriophage contains about 80,000 nucleotide pairs. The recombination study showed that the genetic map contained roughly about 800 recombination units, so that the ratio of recombination probability (at small distances) to molecular distance would be about 0·01 per cent recombination per nucleotide pair. Thus, if two strains having mutations one nucleotide pair apart are crossed, the proportion of recombinants in the progeny should be 0·01 per cent. The smallest non-zero recombination value so far observed among the mutants of T4 bacteriophage is about 0·02 per cent recombination. Benzer[41] denotes this smallest unit as a 'recon', the smallest element in the unidimensional array of genetic units that is interchangeable, but not divisible, by genetic recombination. The size of the recon, therefore, appears to be limited to about two nucleotide pairs. Similarly, Benzer defines the 'muton' as the smallest element, alteration of which can be effective in causing a mutation, and this unit on the same basis would, apparently, involve no more than five nucleotide pairs. The functional genetic unit (the 'cistron') is less readily defined and identified with molecular parameters. All these estimates are very rough but the possibility of their calculation at all shows that the gap between the molecular and cytological levels of organization is being bridged.

REFERENCES

[1] Freese, E. *Cold Spr. Harb. Symp. quant. Biol.*, 1958, **23**, 13
[2] Shooter, K. V. *Progr. Biophys.*, 1957, **8**, 310
[3] Peacocke, A. R. *Progr. Biophys.*, 1960, **10**, 55
[4] Davison, P. F., Conway, B. E. and Butler, J. A. V. *Progr. Biophys.*, 1954, **4**, 148
[5] Bacq, Z. M. and Alexander, P. A. *Fundamentals of Radiobiology.* London; Butterworths, 1955.
[6] Peacocke, A. R. and Pritchard, N. J. 'Biophysical aspects of Ultrasound' in *Progress in Biophysics* Vol. 18 (in the press) Pergamon, Oxford
[6a] Jordan, D. O. *The Chemistry of Nucleic Acids*, Chapter 11. London; Butterworths, 1960
[7] Doty, P., McGill, B. B. and Rice, S. A. *Proc. nat. Acad. Sci., Wash.*, 1958, **44**, 432
[8] Davison, P. F. *Proc. nat. Acad. Sci., Wash.*, 1959, **45**, 1560; *Nature, Lond.*, 1960, **185**, 918
[9] Schmidt, G. In *The Nucleic Acids*, (Ed. Chargaff and Davidson), Chapter 15. London; Academic Press, 1955
[10] Thomas, C. A. *J. Amer. Chem. Soc.*, 1956, **78**, 1861
[11] Lawley, P. D. *J. Chim. phys.*, 1961, **58**, p. 1011
[12] Rice, S. A. and Doty, P., *J. Amer. chem. Soc.*, 1957, **79**, 3937
[13] Tamm, C., Hodes, M. E. and Chargaff, E. *J. biol. Chem.*, 1952, **195**, 49; Thomas, C. A. and Doty, P. *J. Amer. chem. Soc.*, 1956, **78**, 1854
[14] Luzzati, D., Schweitz, H., Bach, M. and Chevallier, M. *J. Chim. phys.*, 1961, **58**, 1021; Latarjet, R., Rebeyrotte, N. and Demerseman, P. In *Les peroxydes organiques en radiobiologie*, p. 61. Paris; Massons et cie, 1958
[15] Litt, M., Marmur, J., Ephrussi-Taylor, H. and Doty, P. *Proc. nat. Acad. Sci., Wash.*, 1958, **44**, 144
[16] Benzer, S. *Proc. nat. Acad. Sci., Wash.*, 1955, **41**, 344
[17] Ambrose, E. J. *Progr. Biophys.*, 1956, **6**, 25
[18] Butler, J. A. V. *Experientia*, 1955, **11**, 289
[19] Berends, W. *J. Chim. phys.*, 1961, **58**, 1034; Wacker, A., Dellweg, H. and Weinblum, D. *Naturwissenschaften*, 1960, **47**, 477; Wacker, A., Dellweg, H. and Lodeman, E. *Angew. Chem.*, 1961, **73**, 64
[20] Schuster, H. and Schramm, G. *Z. Naturf.*, 1958, **13b**, 697; Vielmetter, W. and Schuster, H. *Z. Naturf.*, 1960, **15b**, 304
[21] Schuster, H. and Vielmetter, W. *J. Chim. phys.*, 1961, **58**, 1005.
[22] Litman, R. M. *J. Chim. phys.*, 1961, **58**, 997.

[23] Lerman, L. S. and Tolmach, L. J. *Biochim. biophys. Acta*, 1959, **33**, 371; Litman, R. M. and Ephrussi-Taylor, H., *C.R., Acad. Sci. Paris*, 1959, **249**, 838

[24] Basilio, C., Wahba, A. J., Lengyel, P., Speyer, J. F. and Ochoa, S. *Proc. nat. Acad. Sci., Wash.*, 1962, **48**, 613

[25] Peacocke, A. R. and Walker, I. O. *J. mol. Biol.*, 1962, **5**, 550, 560, 564

[26] Peacocke, A. R. *Chem. Soc. Special Publ.*, 1957, **8**, 139

[27] Thomas, R. *Biochim. biophys. Acta*, 1954, **14**, 231; Cavalieri, L. F., Rosoff, M. and Rosenberg, B. H. *J. Amer. chem. Soc.*, 1956, **78**, 5239

[28] Cox, R. A., Wilson, S., Overend, W. G. and Peacocke, A. R. *Nature, Lond.*, 1955, **176**, 919; Peacocke, A. R. and Preston, B. N. *Proc. Roy. Soc.*, 1960, **153B**, 103

[29] Grossman, L., Stollar, D. and Herrington, K. *J. Chim. phys.*, 1961, **58**, 1078

[30] Zamenhof, S., Alexander, H. E. and Leidy, G., *J. exp. Med.*, 1953, **98**, 373

[31] Doty, P., Marmur, J., Eigner, J. and Schildkraut, C. *Proc. nat. Acad. Sci., Wash.*, 1960, **46**, 461

[32] Marmur, J. and Lane, D. *Proc. nat. Acad. Sci., Wash.*, 1960, **46**, 453

[33] Darlington, C. D. and La Cour, L. F. *The Handling of Chromosomes*. London; Allen and Unwin, 1962

[34] Darlington, C. D. *Nature, Lond.*, 1955, **176**, 1139

[35] Alexander, P. and Stacey, K. A. *Proc. IVth Inter. Congr. Biochem.*, 1958, Vienna, Symposium No. IX, p. 98

[36] Alexander, P. A. and Stacey, K. A. *Progress in Radiobiology*, p. 105. Edinburgh; Oliver and Boyd, 1956

[37] Setlow, R. and Doyle, B. *Biochim. biophys. Acta*, 1954, **15**, 117

[38] Geiduschek, E. P. *Proc. nat. Acad. Sci., Wash.*, 1961, **47**, 950

[39] Fraenkel-Conrat, H. *Biochim. biophys. Acta*, 1961, **49**, 169

[40] Loveless, A. *Proc. Roy. Soc.*, 1959, **150B**, 497

[41] Benzer, S. In *The Chemical Basis of Heredity*, (Ed. McElroy and Glass), p. 70. Baltimore; Johns Hopkins Press, 1957

CHAPTER 10

THE CONTROL OF PROTEIN SYNTHESIS BY NUCLEIC ACIDS

In retrospect the first indication that genes function by controlling enzymes can be seen in the work of Garrod at the beginning of this century. Garrod studied a number of human diseases, particularly alkaptonuria, and suggested that they were 'inborn errors in metabolism'[1]. Alkaptonurics excrete large quantities of 2:5-di-hydroxyphenylacetic acid. Garrod recognized that the excretion of this acid was the result of a block in a normal sequence of biochemical reactions and that the block was the result of a genetic deficiency.

Further indications in the same direction were provided by studies on the chemical structure and genetic control of the flower pigments in a series of Streptocarpus hybrids[2,3]. The pigments involved are anthocyanins differing from each other in the substituents attached to the rings of the parent compound. Pigment inheritance seems to be controlled by three genes each of which can exist in two alternative forms or alleles. The particular alleles present in any plant determine which substituents are present on the anthocyanin ring. Individual genes thus seem to correspond to specific steps in the synthesis of the flower pigments.

Studies by Ephrussi and Beadle on the eye pigments of *Drosophila melanogaster*[4,5] provided more information on the genetic control of biochemical processes. Wild-type Drosophila have dark red eyes because of the presence of two pigments, one red and the other brown. Several genes are known which modify the production of the brown pigment. Flies homozygous for recessive alleles of any of these genes have red eyes because no brown pigment is present. Many of the structures (including eyes) present in adult flies develop from imaginal discs present in the larvae. If eye imaginal discs are transplanted from one larva to the body cavity of another then adult flies with an additional non-functional eye may be obtained. Beadle and Ephrussi carried out a series of experiments in which eye discs from larvae of one genotype were transferred to larvae of another genotype as shown in Table 10. 1. The experiments were interpreted as showing that the lymph of wild-type flies contains compounds which the mutant eye

TABLE 10.1

Genotype of implant	Genotype of host	Phenotype of non-functional eye
v	+	+
cn	+	+
+	v	+
+	cn	+
v	cn	+
cn	v	cn

v = vermilion eye colour   cn = cinnabar eye colour
+ = wild type eye colour

The sequence of control by genes in biosynthesis of Drosophila eye pigments is inferred from the above and other data as:

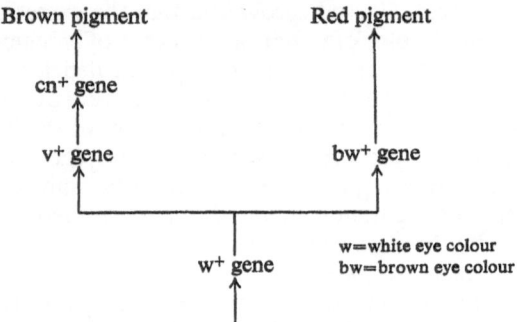

w=white eye colour
bw=brown eye colour

discs do not contain and which are necessary for the synthesis of brown pigment. The implantation of vermilion eye discs into cinnabar larvae and vice versa indicate that two substances are involved, one which changes vermilion to cinnabar and another which converts cinnabar to wild type. These two substances were later identified as kynurenine and hydroxykynurenine, both oxidation products of tryptophan. Mutation of either of the wild-type genes, v^+ or cn^+, appears to result in a block at one step in the oxidation of tryptophan and Ephrussi[6] suggested that the genes involved might control the production of enzymes involved in tryptophan oxidation.

Such early studies of biochemical genetics showed that in suitable systems, mutation of specific genes could lead to a block in the formation of a specific compound but the organisms were unsuitable for more detailed exploration of the mechanism of gene action. Spectacular advances in this direction followed the introduction of microorganisms for use in genetic investigations.

118

Many cases are now known in which mutant organisms are deficient in their intermediary metabolism[7, 8]. In some of these examples, the enzyme necessary for the step which is blocked cannot be demonstrated *in vitro* using cell-free preparations under conditions in which a normal, wild-type strain does show enzyme activity. In other examples the mutant has a much lower concentration of the enzyme or other specific protein than the wild-type strain. Thus there is a correlation between the particular allele of a gene present and the presence of a particular protein. The simplest interpretation of this is that the gene is an essential part of the enzyme-forming mechanism. The possibility that genes may act by determining the specificity of enzymes was suggested by Beadle and Tatum[9] and developed into the 'one gene–one enzyme' hypothesis[5]. In view of more recent work this should now be rephrased as 'one gene–one polypeptide chain'.

Not all gene mutations result in differences in amount of the specific protein of a mutant organism. Numerous cases are now known in which a specific mutation has resulted in a structural change in a particular protein. These are the cases which have revealed the relationship between genes and proteins and shown that one function of genes is to control the sequence of amino acids in proteins.

Horowitz and Fling[10] found that in Neurospora the enzyme tyrosinase which is necessary for the formation of the black pigment melanin could exist in two different forms. Some strains produced the pigment only when grown at temperatures of 25°C or lower while other strains produced pigment at all temperatues at which the organism is capable of growth. Tyrosinase from the two strains differed considerably in their stability to heat; the stable and labile forms had half lives at 59°C of 30 and 3 minutes respectively. A given pure strain of Neurospora produced only one kind of tyrosinase. When two such pure strains, differing with respect to tyrosinase, were crossed, half the progeny resembled one parent and the other half resembled the other. Thus the difference was inherited as if caused by a pair of allelic genes, one determining the production of the stable and the other of the labile form of the enzyme. It is possible in Neurospora to form a heterocaryon, that is, a culture whose cells contain nuclei of different genetic constitution. A heterocaryon containing both alleles of the tyrosinase gene produced both types of tyrosinase. Careful examination showed that the difference between the two forms of the enzyme was a property of the proteins themselves, that is, it was structural, and was not caused by the presence of inhibitors or activators. In this example, the change in protein structure resulting from gene mutation has not been further characterized, but in some other systems a very precise correlation has been achieved between gene mutation and the structural change which has resulted in a specific protein.

The first example in which a correlation was shown between gene mutation and a change in the amino acid sequence of a protein resulted from studies on the biochemistry and genetics of sickle cell anaemia, a disease found in negro populations in Africa and the United States of America. It is generally accepted[11] that the disease is inherited in a simple Mendelian manner, individuals suffering from the disease being homozygous for the abnormal allele of the gene concerned. A characteristic of the disease, which is often fatal before late adolescence, is that under conditions of low oxygen tension the red blood cells change from the normal biconcave shape to filamentous and sickle-shaped forms.

Pauling and colleagues[12] showed that haemoglobin from sickle cell anaemics had a different electrophoretic mobility from normal haemoglobin and afterwards Perutz and Mitchison[13] found that in the reduced condition it was also much less soluble than normal haemoglobin. Sickle cells are probably formed because haemoglobin inside the cell aggregates under conditions of low oxygen tension and thereby distorts the shape of the cell.

The difference in electrophoretic mobility of normal and sickle cell haemoglobin indicates a difference in the net charge of molecules from the two sources. After the haem portions were found to be identical it was shown that the protein components differed in mobility on electrophoresis. Total amino acid analysis failed to show any difference in composition between the proteins from the two types of haemoglobin. In 1956, however, Ingram[14] applied the methods developed by Sanger[15] in his work on the structure of insulin and discovered the nature of the difference between the two proteins.

Ingram examined the peptides produced by tryptic digestion of the proteins. Trypsin hydrolyses only those peptide links which involve lysine or arginine carboxyl groups and as a result it produces a limited number of peptides. The peptides were separated into a pattern of spots on paper ('finger-prints') by a combination of electrophoresis and chromatography. All except one of the peptides obtained occupied identical positions on the two finger-prints. The amino acid sequences of the anomalous peptides were determined and it was found that the peptides differed only in that the sickle cell peptide contained a valine residue in a position occupied by glutamic acid in the peptide from normal haemoglobin (Table 10.2). All the other peptides from the two proteins appeared to have identical sequences.

Another form of haemoglobin (C) has been shown to differ from normal and sickle cell haemoglobin[16] in having a lysine residue replace the same glutamic acid residue which in sickle cell protein is replaced by valine (Table 10.2). This result is particularly remarkable because sickle cell anaemia and haemoglobin C anaemia seem to be determined by alleles of the same gene.

Thus gene mutations at a particular locus have resulted in specific amino acid substitutions at a particular site in a protein molecule, which supports the idea that there is a precise relationship between the gene and the position of the amino acid residues in a polypeptide. This evidence however must be accepted with caution because the genetic evidence which is based on human pedigrees lacks the precision which is possible in experiments with other organisms.

TABLE 10.2

Structure of Peptide 4 from Haemoglobins

HbA	Val–His–Leu–Thr–Pro–Glu–Glu–Lys
HbS	„ „ „ „ „ –Val– „ „
HbC	„ „ „ „ „ –Lys– „ „

HbA = normal haemoglobin   HbS = sickle cell haemoglobin
HbC = haemoglobin C
Amino acid notation: *see* page 139

Perhaps the most extensive genetic and biochemical information is reported for the tryptophan synthetase system studied by Yanofsky and his co-workers[17], first of all with Neurospora and then more recently with *E. coli*. This enzyme catalyses the combination of indole and serine to form tryptophan. *In vitro* the enzyme can catalyse the following reactions:

(1) indoleglycerol phosphate → indole + triose phosphate
(2) indole + L-serine → L-tryptophan
(3) indoleglycerol   phosphate + L-serine → L-tryptophan + triose phosphate.

Reaction (3) is not simply the sum of reactions (1) and (2) since it has been shown[18] that indole is not an intermediate in reaction (3). In Neurospora a single protein catalyses all three reactions whereas in *E. coli* the tryptophan synthetase system consists of two separable proteins, A and B. Twenty-five Neurospora mutants which were unable to form tryptophan from indole and serine were studied initially[19]. Genetic tests indicated that all the mutations had occurred at a single gene locus, the *td* locus. Tryptophan synthetase activity could not be demonstrated in extracts of the mutants under conditions in which 1 per cent of the activity of wild-type extracts could have been detected. There was one exception to this. A temperature sensitive strain, *td*-24, was shown to have some enzyme activity but the enzyme was more susceptible to inhibition by heavy metal ions than the wild-type enzyme. A number of the mutant strains produced an antigenic material related to tryptophan synthetase. This cross-reacting material, though lacking enzyme activity, must have been structurally similar to the enzyme protein. One of the cross-reacting materials could in fact catalyse the indoleglycerol phosphate → indole + triose phosphate reaction. These results with Neurospora showed that as a result of mutation a protein with altered properties was formed, but they gave no direct evidence as to the nature of the underlying structural alteration.

Yanofsky's studies with *E. coli* have resulted in a correlation of gene mutation with specific amino acids at specific sites within the A protein of the tryptophan synthetase system. Two main types of A mutants have been found on the basis of enzymic and serological tests. One type forms an altered A protein, the other does not form any protein detectable by these methods. The mutant A proteins differ from wild-type A protein and among themselves in regard to heat stability and stability at low pH. Such differences appear to be characteristic of the mutant proteins themselves and are not due to other substances present in the preparations[20].

To obtain information about the precise chemical nature of the modification induced in the protein as a result of mutation, it was necessary to determine the amino acid sequence of peptides obtained from it. Fortunately the A protein had a fairly low molecular weight of 29,500 and proved amenable to more detailed analysis using the methods developed originally by Sanger and Smith[15] for insulin and Ingram [16] for haemoglobin.

Tryptophan synthetase A protein from wild-type strains gave twenty-five major peptides[20]. The 'finger-prints' of the altered A proteins from a number of mutants differed from that of the wild type. It should be pointed out that not all amino acid substitutions were expected to result in altered peptide patterns since not all amino acid substitutions cause differences in the mobility of peptides under the conditions used. The A protein of two mutants, A23 and A46, which mapped at the same site on the genetic map, were found to differ from normal A protein in the composition of a particular peptide. In the protein from A23 an arginine residue replaced a glycine present in wild-type A protein while in A46 this same glycine residue was replaced by glutamic acid. Several revertants, that is, strains which have regained the ability to grow on minimal medium as a result of mutation, have been revovered from each of the strains and where these map at the same genetic site it has been shown that the amino acids substituted as a result of mutation occupy the same position in the corresponding protein[21] (Table 10.3). The results provide evidence that eight independent mutations, mapping at the same site, have resulted in the substitution of six different amino acids at the same position in the polypeptide chain of tryptophan synthetase A protein.

TABLE 10.3

Amino Acid Substitutions in Tryptic Peptide TP3C1 from *E. coli* Tryptophan Synthetase A Proteins[21]

Strain	Amino acid sequence
A46–PR9*	Val
A46–FR2†	Gly
A46–FR1	Ala
A46	Glu
Wild type	AspN–Ala–Ala–Pro–Pro–Leu–GluN–Gly–Phe
A23	Arg
A23–FR1	Gly
A23–FR2	Ser

Amino acid notation: page 139

*PR = Partial Revertant
†FR = Full Revertant

With TMV (Chapter 3 and Chapter 12), the particles of which contain only RNA and protein, it appears that the sequence of amino acids in the coat protein is determined by the RNA. The coat protein contains 158 amino acid residues and the complete sequence of the residues in the polypeptide is known[22]. After the amino acid sequence of the wild-type virus was established a number of mutants were examined and differences in the coat proteins identified. This work showed that mutations which affect the protein of the virus particles do so by causing single amino acid substitutions[23, 24] (Table 12.2) Recombinational analysis of mutational sites is not yet possible in TMV, so that it is not possible to correlate the amino acid substitutions with a genetic map in this organism.

Infection of bacterial cells by an intemperate bacteriophage causes profound changes in the metabolism of the host for the phage appears to redirect the biosynthetic machinery of the host cell to the task of phage formation. Bacterial DNA does not contain 5-hydroxymethyl cytosine, which is a constituent of the DNA of T-even phages, and the bacterial cell lacks some of the enzymes necessary for its synthesis[25]. After infection, however, the necessary enzymes are synthesized under the control of viral DNA. There is a reasonable amount of evidence[26] that these enzymes are synthesized under the control of structural genes in the phage itself.

So far only the evidence concerning the function of structural genes, that is, genes which specify the amino acid sequence of an enzyme or other protein, has been considered. In addition to this, evidence is available that, at least in bacteria, genes with regulating or controlling functions exist in two types[27]; (a) regulator genes in which mutations affect the conditions and rate of synthesis of the corresponding protein and (b) operator genes, which control the rate of transcription of one or more structural genes adjacent to them in the chromosome.

Marked changes in the enzymic content of cells usually involve either repression or induction of enzyme synthesis. Repression of synthesis of enzymes involved in a biosynthetic pathway may be caused by the addition of the end product of the pathway to a bacterial culture[28]. In enzyme induction, specific enzymes are produced by cells as a result of the exposure of the cells to specific substrates. Both induction and repression probably have the same basic mechanism and in both systems mutations affecting the regulator genes are known.

Mutations resulting in uncontrolled synthesis of specific proteins irrespective of the presence or absence of the compound by which the wild-type strain is repressed or induced have been obtained in both systems[29, 30]. These regulator gene mutations are found invariably to map at sites distinct from those of the structural gene or genes governing the protein or proteins concerned. Moreover, the modified regulator genes do not appear to alter the structure of the proteins produced.

Evidence regarding the mode of action of regulator genes has been obtained from experiments with cells heterozygous for a normal and a mutant regulator gene. Such experiments[31] show that a *single* regulator gene controls the expression of the structural genes in *both* chromosomes. These results are most simply explained by the assumption that the regulator gene produces a specific cytoplasmic product, the repressor, which inhibits protein synthesis. Mutation in the regulator gene, on this interpretation, results in the production of an inactive repressor. Direct identification of the repressor is not yet possible but several experiments[32] indicate that the repressor probably is, or involves, a protein.

The concept of the operator gene as a unit which controls the functioning of an adjacent chromosomal segment, containing one or more structural genes each of which determines the amino acid sequences of a protein, was suggested by Jacob and Monod[27, 33]. If the existence of a specific repressor is considered as established then the existence of an operator as its site of action becomes essential.

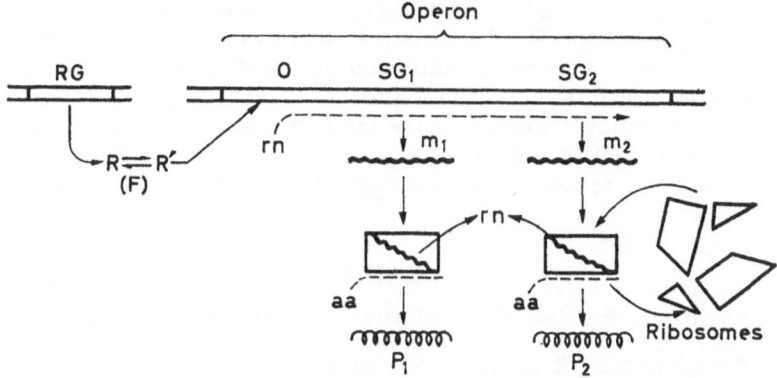

Figure 10.1. General model[36] of the regulation of enzyme synthesis. RG: regulator gene; R: repressor converted to R' in presence of effector F (inducing or repressing metabolite); O: operator; SG_1, SG_2: structural genes; rn: ribonucleotides; m_1, m_2: messengers made by SG_1 and SG_2; aa: amino acids; P_1, P_2: proteins made by ribosomes associated with m_1 and m_2

(Reproduced by courtesy of the Editor of *Cold Spring Harbor Symposia on Quantitative Biology*)

Operator mutations are known in a number of bacterial systems[34]. Experiments with strains heterozygous for wild-type and mutant operator genes indicate that, unlike regulator genes, the operator gene does not act via a cytoplasmic product but controls directly the transcription of the adjacent chromosomal segment containing the structural genes. The properties of operator mutants indicate that the function of the normal operator gene is to act as the receptor of a specific repressor.

In Salmonella nine genes controlling eight enzymes involved in histidine biosynthesis occupy adjacent sites in the bacterial chromosome. Small deletions affecting a gene located at one extremity of the group result in the loss by the cells of the ability to produce all eight enzymes. Recombination experiments indicate that the structural genes for these enzymes are present and normal in the strains carrying the deletions. This sequence of genes, therefore, constitutes an 'operon' controlled by a single operator located at one side of it.

According to the operon theory, if a structural gene is removed from the operon and incorporated elsewhere in the genome as a result of a chromosomal rearrangement then the structural gene should escape from the control of the operator. Using the Salmonella histidine system this prediction has been tested and found to be correct[35].

The general picture of the control of protein biosynthesis which emerges from the study of bacterial systems is one in which regulator genes transmit specific signals (repressors) to operator genes which regulate the rate of information transfer (that is, RNA synthesis) from the structural genes to the protein-forming centres of the cell. The information contained in the DNA is regarded as being transcribed in segments which correspond to operons, the transcription starting from specific points, the operator genes[36] (see *Figure 10.1*).

REFERENCES

[1] Garrod, A. E. *Inborn Errors of Metabolism.* Oxford; University Press, 1923
[2] Scott-Moncrieff, R. *J. Genet.*, 1936, **32**, 117
[3] Lawrence, W. J. C. *Biochem. Soc. Symp.*, 1950, **4**, 3
[4] Ephrussi, B. and Beadle, G. W. *Biol. Bull.*, 1937, **71**, 75
[5] Beadle, G. W. *Chem. Rev.*, 1945, **37**, 15
[6] Ephrussi, B. *Cold Spr. Harb. Symp. quant. Biol.*, 1942, **10**, 40
[7] Harris, H. *Human Biochemical Genetics.* Cambridge; University Press, 1959
[8] De Busk, A. G. *Advanc. Enzymol.*, 1956, **17**, 393
[9] Beadle, G. W. and Tatum, E. L. *Proc. nat. Acad. Sci., Wash.*, 1941, **27**, 499
[10] Horowitz, N. H. and Fling, M. In *Enzymes: Units of Biological Structure and Function*, (Ed. Gaebler), p. 139. New York; Academic Press, 1956
[11] Neel, J. V. *Proc. 10th Inter. Cong. Genet.*, 1958, 108
[12] Pauling, L., Itano, H. A., Singer, S. J. and Wells, I. C. *Nature, Lond.*, 1950, **166**, 677
[13] Perutz, M. F. and Mitchison, J. M. *Nature, Lond.*, 1950, **166**, 677
[14] Ingram, V. M. *Nature, Lond.*, 1956, **178**, 792
[15] Sanger, F. and Smith, L. F. *Endeavour*, 1957, **16**, 48
[16] Ingram, V. M. *Hemoglobin and its Abnormalities.* Springfield, Ill.; Charles C. Thomas, 1960
[17] Yanofsky, C. *Bact. Rev.*, 1961, **24**, 221
[18] Yanofsky, C. and Rachmeler, M. *Biochim. biophys. Acta*, 1958, **28**, 640
[19] Yanofsky, C. *Enzymes: Units of Biological Structure and Function* (Ed. Gaebler), p. 147. New York; Academic Press, 1956

REFERENCES

[20] Yanofsky, C. In *Methodology in Basic Genetics*, (Ed. Burdette), p. 303–317. San Francisco; Holden-Day Inc., 1963

[21] Carlton, B. C. and Yanofsky, C. *J. biol. Chem.*, 1963, **238**, 2390

[22] Tsugita, A., Gish, D. T., Young, J., Fraenkel-Conrat, H., Knight, C. A. and Stanley, W. M. *Proc. nat. Acad. Sci.*, *Wash.*, 1960, **46**, 1463

[23] Wittmann, H. G. *Naturwissenschaften*, 1961, **48**, 729

[24] Tsugita, A. and Fraenkel-Conrat, H. *J. mol. Biol.*, 1962, **4**, 73

[25] Cohen, S. S. In *Growth in Living Systems*, (Ed. Zarrow), p. 17–38. New York; Basic Books Inc., 1961

[26] Bessman, M. J. *J. biol. Chem.*, 1959, **234**, 2735; Thomas, C. A. and Suskind, S. R. *Virology*, 1960, **12**, 1; Aposhian, H. V. and Kornberg, A. *Fed. Proc.*, 1961, **20**, 361

[27] Jacob, F. and Monod, J. *J. mol. Biol.*, 1961, **3**, 318

[28] Gorini, L. and Maas, W. K. In *The Chemical Basis of Heredity*, (Ed. McElroy and Glass), p. 469. Johns Hopkins Press; Baltimore, 1958; Ames, B. N. and Garry, B. *Proc. nat. Acad. Sci.*, *Wash.*, 1959, **45**, 1453

[29] Cohen, G. N. and Jacob, F. *C. R. Acad. Sci.*, 1959, **248**, 3490; Gorini, L., Gundersen, W. and Burger, M. *Cold Spr. Harb. Symp. quant. Biol.*, 1961, **26**, 173

[30] Kogut, M., Pollock, M. and Tridgell, E. J. *Biochem. J.*, 1956, **62**, 391; Pardee, A. B., Jacob, F. and Monod, J. *J. mol. Biol.*, 1959, **1**, 165; Kalckar, H. and Sundararajan, T. A. *Cold Spr. Harb. Symp. quant. Biol.*, 1961, **26**, 227

[31] Pardee, A. B., Jacob, F. and Monod, J. *J. mol. Biol.*, 1959, **1**, 165; Horiuchi, T., Horiuchi, S. and Novick, A. *J. mol. Biol.*, 1961, **3**, 703; Sussman, R. and Jacob, F. *C. R. Acad. Sci.*, 1962, **254**, 1517

[32] Yanofsky, C., Helinski, D. R., and Maling, B. D. *Cold Spr. Harb. Symp. quant. Biol.*, 1961, **26**, 11; Benzer, S. and Champe, S. P. *Proc. nat. Acad. Sci.*, *Wash.*, 1961, **47**, 1025; Jacob, F., Sussman, R. and Monod, J. *C. R. Acad. Sci.*, 1962, **254**, 4214

[33] Jacob, F. and Monod, J. *C. R. Acad. Sci.*, 1959, **249**, 1282

[34] Ames, B. N., Garry, B. and Herzenberg, L. A. *J. gen. Microbiol.*, 1960, **22**, 369; Lee, N. and Englesberg, E. *Proc. nat. Acad. Sci.*, *Wash.*, 1962, **48**, 335

[35] Ames, B. N., Hartman, P. E. and Jacob, F. *J. mol. Biol.*, 1963, **7**, 23

[36] Jacob, F. and Monod, J. *Cold Spr. Harb. Symp. quant. Biol.*, 1961, **26**, 193

CHAPTER 11

THE MECHANISM OF PROTEIN SYNTHESIS

Although the synthesis of proteins is genetically controlled (Chapter 10) it has been known for some time that most proteins are synthesized in the cytoplasm and not in the nucleus. The generally accepted scheme of protein synthesis is that the actual site of synthesis is the ribosome, a cytoplasmic particle. Ribosomes synthesize proteins only when combined with messenger RNA, which is thought to be the primary product of the structural genes and to act as a template ensuring the correct sequence of amino acids in the protein being synthesized.

The amino acids necessary for the synthesis of the protein reach the ribosomes attached to different transfer RNAs, which are thought to contain a coding segment allowing them to recognize and pair with the corresponding segments of messenger RNA. The amino acids are thus aligned in the correct sequence for the specific protein being formed.

The reaction consists of the following main steps.

(1) AA + ATP + E → (Aacyl AMP-E) + PP
(2) (AacylAMP-E) + S-RNA → (Aacyl S-RNA) + AMP + E
(3) (Aacyl S-RNA) + mRNA + Ribosomes → Protein + S-RNA + mRNA
                                                                    + Ribosomes

AA = Amino acid;   ATP = adenosine triphosphate;   AMP = adenosine monophosphate; E = enzyme; AacylAMP = aminoacyl AMP (Formula XVI); PP = pyrophosphate;   S-RNA = soluble, transfer RNA;   Aacyl S-RNA = compound of amino acid and S-RNA (Formula XVI); mRNA = messenger RNA.

In this chapter the evidence for this scheme is discussed.

SITE OF PROTEIN SYNTHESIS

Circumstantial evidence implicating RNA in protein synthesis was obtained by Caspersson[1] and by Brachet[2]. Both workers concluded that RNA is localized mainly in the nucleolus and in the cytoplasm and that there was a correlation between the RNA content of cells and their ability to synthesize protein. Rapidly dividing cells, such as are found in root tips and embryonic tissue, and protein-secreting cells contain a relatively high concentration of RNA, whereas much less RNA is present in cells not actively engaged in protein biosynthesis.

Further evidence of this type has been obtained using enucleate algae, amoeba and reticulocytes. In the green alga *Acetabularia mediterranea*, the rate of incorporation of isotopically labelled glycine or carbon dioxide remains constant for some time following sectioning of the organism[3]. Vanderhaeghe[4] demonstrated that net protein synthesis may continue for 2 weeks in both nucleated and non-nucleated halves of this alga. RNA synthesis also continues in both nucleated and enucleated halves of Acetabularia[5] indicating that cytoplasmic RNA, as well as protein, may be synthesized in the absence of the nucleus, but after about 10 days the synthesis of both types of macromolecule decreases. Enucleate fragments of *Amoeba proteus* gave different results, perhaps because such fragments were unable to catch prey; in such fragments the RNA content and the ability to incorporate amino acids into protein decreased[6]. During the life cycle of red blood cells a correspondence exists between the RNA content and the ability of the cell to form protein. Immature red blood cells lack a nucleus but contain RNA and are able to synthesize both RNA and protein, but mature cells contain little RNA and have virtually no capacity for protein synthesis[7]. The evidence quoted so far indicates that protein synthesis can occur in the cytoplasm in the absence of the nucleus and that RNA is probably involved. Evidence that the nucleus may synthesize protein has been obtained by the demonstration that isolated thymus nuclei were able to incorporate amino acids into their proteins *in vitro*[8].

Recent confirmation of this picture has been obtained by Sirlin[9] and by Zalokar[10] using autoradiography to localize the site of protein formation following the injection of ³H-leucine into Smittia and Drosophila, respectively. In the salivary tissue of the chironomid the cytoplasm becomes labelled more rapidly and remains more heavily labelled than the nucleus; this indicates local synthesis of protein in the cytoplasm. The results with Drosophila are similar, although in this organism the nuclei ultimately become as heavily labelled as the cytoplasm.

The use of cell-free and microbial systems has permitted the precise localization of most protein synthesis in the ribonucleic acid fraction of the microsomes[11].

THE FUNCTION OF SOLUBLE OR TRANSFER RNA (S-RNA)

The initial stage in protein biosynthesis appears to be the activation of amino acids by activating enzymes in the presence of ATP[12]. These enzymes catalyse the reactions (1) and (2), page 128, and as a result, the amino acid becomes attached to S-RNA. In the first reaction the terminal diphosphate group of ATP is displaced by the α-carboxyl group of the amino acid and an aminoacyl adenylate is formed (XVI). The aminoacyl adenylates are strongly bound to the enzyme and indeed no free adenylate can be detected[13], although adenylate bound to enzyme has been isolated when large amounts of tryptophan-activating enzyme have been used[14]. Combination with the activating enzyme probably prevents the aminoacyl adenylates from reacting at random to form polypeptides[15]. Evidence that the route of synthesis of the activated amino acid is as indicated in (1) and (2) has been obtained from experiments in which the transfer of ^{18}O between the amino acid carboxyl group and AMP has been followed.

The second reaction (2) in which the activating enzymes participate has been demonstrated directly using the complex formed in the first part of the reaction but it is normally measured by following the incorporation of ^{14}C-labelled amino acid into S-RNA in the overall reaction. Attempts to separate the two activities of activating enzymes

XVI

during purification have failed[16], so the present assumption is that the same enzyme catalyses the two reactions.

Soon after the discovery of amino acid activating enzymes their specificity for individual amino acids was shown for, if the total concentration of amino acids was kept constant, reaction (1) was enhanced as the number of different amino acids increased, which suggested that different amino acids did not compete for the same enzyme. About ten amino acid activating enzymes have now been purified and their substrate specificity determined. Most of these activate only a single amino acid[17], but in some cases[18] analogues of an amino acid may also react with the enzyme. The isoleucine activating enzyme is an exception to the general rule as it will also activate valine[19] but it does not transfer valine to S-RNA. Thus, in addition to a specificity for a particular amino acid, the activating enzymes or the aminoacyladenylate-enzyme complexes have a specificity towards S-RNA molecules. A species specificity is also shown by the activating enzymes which vary in their affinity for S-RNA from different species, for example, *E. coli* activating enzymes do not attach arginine to S-RNA from yeast[20].

After the discovery that RNA from cell-free rat liver preparations[21] could bind amino acids, RNAs with similar properties were found in other animal tissues and in micro-organisms. Fractionation of the RNA from such systems showed that the RNA maximally labelled by the amino acid occurred in the supernatant and led to the name soluble RNA (S-RNA). It has a molecular weight of about 25,000[22] and it is unlikely to be either a breakdown product or a precursor of larger molecular weight RNA, but it is uncertain where or how it is synthesized. S-RNA has been discussed more fully in Chapter 5, p. 44 ff.

All S-RNAs have the same end groupings with guanosine phosphate[22] at one end and the sequence cytidylic-cytidylic-adenosine[23,24] at the other. This latter end group shows a rapid turnover and is the point of attachment of the amino acid when the aminoacyl linkage to S-RNA is formed. The amino acid is joined to the terminal adenosine of S-RNA by means of an ester link involving the carboxyl group of the amino acid and the 2' or 3'-hydroxyl group of the ribose of the adenosine[25] (XVI). Cleavage of this ester bond may provide the free energy for the formation of the peptide linkage.

Although all S-RNA molecules have the same terminal sequences there is evidence that there are specific S-RNA sequences for individual amino acids, just as there are activating enzymes specific for individual amino acids. The first evidence for amino acid specific S-RNA fractions came from experiments in which it was shown that there was no competition among various amino acids for binding sites on S-RNA. This approach was elaborated by Berg and co-workers[26] who incubated S-RNA with valine thus preparing valyl-S-RNA. The RNA fraction was then treated with periodate which oxidizes the *cis*-hydroxyl groups on the ribose of the terminal adenosine. The valine was then removed by mild hydrolysis and samples of the S-RNA incubated with various amino acids. Only valine was bound to this S-RNA showing that it was protected by valine from periodate oxidation.

Further evidence for specificity of S-RNA has been obtained from attempts to fractionate it by chromatography[27,28]. Fractions specific for particular amino acids have now been obtained and their structures are discussed in Chapter 5. Proof that S-RNA is an intermediate in protein biosynthesis is based largely on kinetic experiments. It is labelled more rapidly than ribosomes by radioactive amino acids[29] and once labelled will transfer its amino acid to the ribosomal particles[30].

The discovery of S-RNA was anticipated by Crick[31] in his 'adaptor hypothesis'. It was known that proteins were actually synthesized on the ribosomes (*see below*) and it was assumed that the

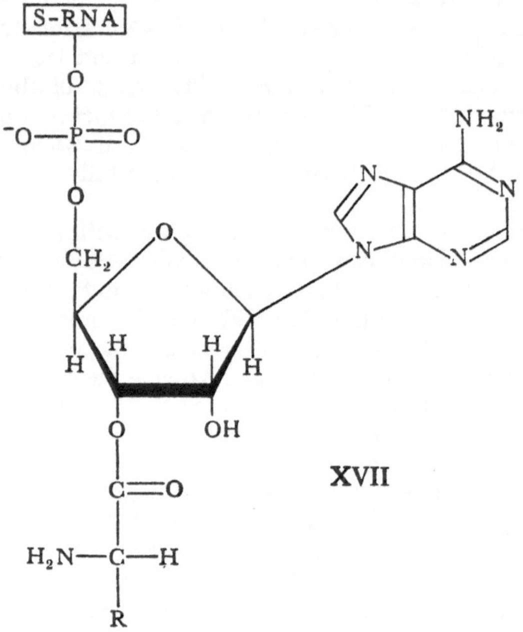

XVII

sequence of the amino acids was determined by the sequence of nucleic acid bases in the ribosomal RNA but the lack of stereochemical complementarity between nucleotides and amino acids made it difficult to see how the RNA could act as a template for protein formation. Crick overcame this difficulty by proposing that the amino acids do not actually combine with the template but that each combines with a specific adaptor molecule which is responsible for locating the amino acid in the correct position on the particle RNA. Presumably correct positioning of the adaptor molecule is due to hydrogen bonding between the coding unit (*see* Chapter 12) of the ribosomal template and the complementary segment of the S-RNA. This scheme requires at least 20 different adaptor molecules, each specific for one of the amino acids found in proteins.

As has been indicated S-RNAs differ in their specificity and pure

132

S-RNA fractions specific for individual amino acids have been obtained. If the adaptor hypothesis (*see* page 44ff) is correct these S-RNAs should differ in at least part of their base sequences. Marked differences have in fact been described in the sequence of bases in different S-RNA fractions[32] (Chapter 5).

RIBOSOMES

The term microsome was used originally to describe a fraction obtained by differential centrifugation of disrupted cells. The microsomal fraction was obtained as a pellet by centrifuging at 100,000 g for two hours. Electron microscopy indicated that the pellet contained mainly membranes to which dense particles were attached. Treatment with deoxycholate dissolves the membranes leaving a fraction containing the particles. These particles are about 200 Å in diameter and contain practically all the microsomal RNA. The term ribosome is used to describe these particles.

Zamecnik and his co-workers[33] identified the site of protein synthesis by showing that the ribosomes are the first part of the cell to incorporate leucine into their proteins. In their experiment a saturating dose of ^{14}C-leucine was injected into a rat and liver lobes were removed successively. The specific activity of various cell fractions was examined and it was found that the ribosome fraction was labelled most rapidly reaching maximum specific activity in about three minutes, after which its activity remained constant. Incorporation of radioactivity into two other cell protein fractions continued at a linear rate throughout the experiment. These results are consistent with the hypothesis that the peptide links of proteins are formed on the ribosomes and that the proteins thus synthesized are subsequently released from the particles and appear in the other fractions examined. Essentially similar results have been obtained by numerous other workers[34] using both whole cell and cell-free systems.

If S-RNA has an adaptor function it must form a definite, if transient, association with the ribosomes. Such an association has been demonstrated[35] in several cases using ^{32}P-labelled S-RNA.

In the early experiments on protein synthesis with ribosomes it was assumed that the ribosomal RNA was the template for protein synthesis. Each gene was thought to determine the base sequence of one kind of ribosomal RNA which then combined with protein to form morphologically similar, but genetically distinct, ribosomes. Several lines of evidence now indicate that ribosomal RNA does not, in fact, form the template for protein biosynthesis. The fate of radioactive precursors incorporated into ribosomal RNA showed that the

.ribosomal RNA was stable. Later more elaborate experiments[36], in which heavy isotopes of carbon and nitrogen were used, confirmed this suggestion.

Evidence then began to accumulate that the template was unstable and this cast doubt on the suggested role for ribosomal RNA in protein biosynthesis. Experiments with the β-galactosidase system of *E. coli* provided the first evidence that the template was unstable[37]. If *E. coli* chromosomes are inactivated by decay of ^{32}P incorporated into them, enzyme formation stops in a few minutes suggesting that the templates cannot function unless DNA is also functioning. In experiments of the opposite type, genes from 'male' cells can be introduced into 'female' cells which will then start to produce enzymes previously missing from the latter. If new ribosomes are necessary for the synthesis of these enzymes then it is expected that synthesis of the new enzymes should start slowly and gradually reach a maximum. Maximum rate of synthesis is attained in 2–3 min after gene transfer[33].

Fluorouracil is an analogue of uracil which can be incorporated into RNA. Once incorporated, fluorouracil is more likely to pair with guanine than with adenine which is the normal complementary base to uracil. When bacteria were grown in the presence of fluorouracil abnormal protein was synthesized and the production of normal protein stopped within a few minutes[38]. The persistence of stable templates does not accord with these observations.

A final point against ribosomal RNA being the template is that its base composition is neither identical with nor complementary to that of DNA. A correspondence between ribosomal RNA and DNA would be expected if this RNA functioned as a template.

Jacob and Monod[39] pointed out that these observations were consistent with the production of an unstable RNA intermediate which carries the genetic message from the gene to the ribosomes and breaks down after only a few protein molecules have been produced on it. They called this RNA 'messenger RNA'.

THE FUNCTION OF MESSENGER RNA (mRNA)

When *E. coli* cells are infected by the phage T2, net synthesis of RNA stops but a small fraction of the RNA turns over rapidly, that is, forms rapidly but breaks down at a fast rate[40]. Volkin and Astrachan[41] confirmed these results and, more importantly, showed that the base composition of the RNA with the fast turnover resembled that of the DNA of the infecting phage, whereas the base composition of the bulk of the RNA did not. This finding suggested that the RNA fraction with the rapid turnover was synthesized on T2 DNA templates. The authors suggested that this RNA might then function as additional templates for phage DNA synthesis inside the host cell. In 1960, however, it was shown[42] that the newly formed RNA was associated with the ribosomes and therefore was unlikely to be concerned in DNA replication. Normura, Hall and Spiegelman[42] suggested that this RNA might form the template for the synthesis of viral protein on ribosomes formed following infection. It is now generally accepted that this RNA is the metabolically unstable template on which protein is synthesized but the suggestion that it combines with ribosomes formed after infection is probably untrue. Two sets of experiments placed this theory on a sound footing.

It was argued[43] that the rapidly synthesized RNA fraction should be associated with ribosomes synthesized before infection and Brenner, Jacob and Meselson designed experiments to test this hypothesis[43]. For the crucial experiment bacteria were grown in medium in which the carbon and nitrogen compounds were enriched with heavy carbon (^{13}C) and heavy nitrogen (^{15}N), respectively. The cells were transferred to normal medium and phage added. Two minutes after infection ^{32}P-phosphate was added to the medium and after a short interval the cells were harvested and ribosomes prepared from them. Density gradient centrifugation (*see* Chapter 8) showed that the radioactivity, and hence the newly synthesized RNA, was associated with the ribosomes made before infection. To complete this scheme of protein synthesis it was necessary to show that protein is actually synthesized on the ribosomes to which the newly formed messenger is attached. This was achieved by growing the bacteria in 'heavy' medium transferring them to normal medium and, after infection, adding ^{35}S-labelled compounds to the medium. The ^{35}S labels protein synthesized after infection of the cells. Centrifugation in a density gradient revealed that the newly synthesized protein, like the messenger RNA, was associated with ribosomes formed before infection.

Meanwhile a similar unstable RNA fraction differing in base composition from ribosomal RNA but analogous to DNA was found in normal uninfected bacteria[44] suggesting that the mode of protein synthesis first demonstrated in phage-infected cells is the normal mode of protein formation. Evidence that mRNA is synthesized on a DNA template was given by Hall and Spiegelman[45] who formed hybrids between T2–DNA and T2–mRNA showing that there is a correspondence between the base sequence of the DNA and the mRNA. (*See* pages 37 and 58.)

FORMATION OF THE POLYPEPTIDE CHAIN

Information about the sequence of peptide bond formation has been obtained from experiments in which the distribution of ^{14}C-amino acids in protein chains has been studied.

Yoshida and Tobita[46] added ^{14}C-leucine to a *B. subtilis* culture and isolated the extracellular amylase produced. Leucine was isolated from the carboxyl end, the amino end and intermediate positions of the enzyme protein and its specific activity measured. The order of decreasing activity of the leucine was: carboxyl end, intermediate positions, amino end. This suggests an ordered synthesis of the polypeptide starting from the amino end, assuming that some polypeptide chains were already initiated when the ^{14}C-leucine was added.

The experiments of Dintzis[47] prove that proteins are synthesized by the step-wise addition of single amino acids beginning with the amino-terminal acid. Dintzis used rabbit reticulocytes which will synthesize haemoglobin when incubated in a nutrient medium. He argued that if protein synthesis proceeded in a step-wise fashion then at any instant the ribosomes should contain chains at various stages of formation which on completion would be released into solution in the cell. If the chains were assembled in sequence, then only the growing end should be labelled if a labelled amino acid is added to the medium. Tritium-labelled leucine was added to a suspension of reticulocytes and after a further period of incubation the haemoglobin was

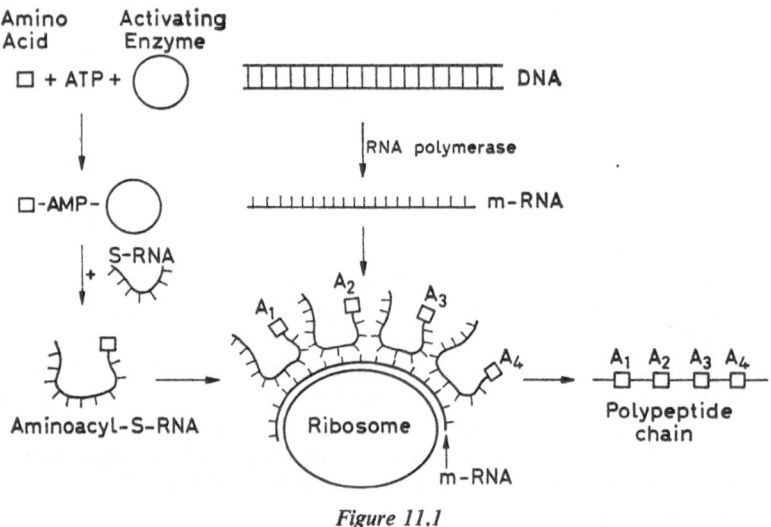

Figure 11.1

extracted, separated into the α and β chains and tryptic digests pre-pared. The tryptic peptides were separated and their specific activity determined.

In agreement with expectation the peptides obtained could be arranged in a linear order according to the specific activity of the leucine they contained. The carboxyl terminal peptide was most active and the amino terminal peptide least active. It has since been shown that the sequence of the tryptic peptides corresponds exactly to the linear order determined by means of the specific activity measure-ments. These experiments thus support the conclusion that proteins are synthesized in sequence starting from the amino end.

The experiments also suggest that the genetic code (Chapter 12) can be transcribed from only one point which corresponds to the amino end of the protein concerned. These ideas on the mechanism of protein synthesis are summarized in *Figure 11.1*.

REFERENCES

[1] Caspersson, T. O. *Naturwissenschaften*, 1941, **29**, 33

[2] Brachet, J. *Arch. Biol.*, *Paris*, 1942, **53**, 207

[3] Brachet, J. and Chantrenne, H. *Nature, Lond.*, 1951, **168**, 950

[4] Vanderhaeghe, F. *Biochim. biophys. Acta*, 1954, **15**, 281

[5] Brachet, J., Chantrenne, H. and Vanderhaeghe, F. *Biochim. biophys. Acta*, 1955, **18**, 544

[6] Linet, N. and Brachet, J., *Biochim. biophys. Acta*, 1951, **7**, 607

[7] Koritz, S. B. and Chantrenne, H., *Biochim. biophys. Acta*, 1954, **13**, 209; London, I. M., Shemin, D. and Rittenberg, D. *J. biol. Chem.*, 1950, **183**, 749

[8] Allfrey, V. G., Mirsky, A. E. and Osawa, S. *J. gen. Physiol.*, 1957, **40**, 451; Allfrey, V. G. and Mirsky, A. E. In *Subcellular Particles*, (Ed. Teru Hayashi), p. 186. New York; Ronald Press, 1959

[9] Sirlin, J. L. *Exp. Cell. Res.*, 1960, **19**, 177

[10] Zalokar, M., *Exp. Cell Res.*, 1960, **19**, 184

[11] Stulberg, M. P. and Novelli, G. D. In *The Enzymes VI*, (Ed. Boyer, Lardy and Myrbäck), p. 401. New York; Academic Press, 1962

[12] Lipmann, F. *Proc. nat. Acad. Sci.*, *Wash.*, 1958, **44**, 67

[13] Hoagland, M. B., Keller, E. B. and Zamecnik, P. C. *J. biol. Chem.*, 1956, **218**, 345

[14] Kingdon, H. S., Webster, L. T. and Davies, E. W. *Proc. nat. Acad. Sci.*, *Wash.*, 1958, **44**, 757

[15] Moldave, K. Castelfranco, P. and Meister, A., *J. biol. Chem.*, 1959, **234**, 841

[16] Berg, P. and Ofengand, E. *Proc. nat. Acad. Sci.*, *Wash.*, 1958, **44**, 78; Holley, R. W. and Goldstein, J. *J. biol. Chem.*, 1959, **234**, 1765

[17] Berg, P. *Ann. Rev. Biochem.*, 1961, **30**, 293

[18] Sharon, N. and Lipmann, F. *Arch. Biochem.*, 1957, **69**, 217

MECHANISM OF PROTEIN SYNTHESIS

[19] Bergmann, F. H., Berg, P. and Dieckmann, M. *J. biol. Chem.*, 1961, **236**, 1735
[20] Benzer, S. and Weisblum, B. *Proc. nat. Acad. Sci., Wash.*, 1961, **47**, 1149
[21] Hoagland, M. B., Zamecnik, P. C. and Stephenson, M. L. *Biochim. biophys. Acta*, 1957, **24**, 215
[22] Otaka, E. and Osawa, S. *Nature, Lond.*, 1960, **175**, 921
[23] Singer, M. and Cantoni, G. *Biochim. biophys. Acta*, 1960, **39**, 182
[24] Canellakis, E. S. *Biochim. biophys. Acta*, 1957, **25**, 217
[25] Zachau, H. G., Acs, G. and Lipmann, F. *Proc. nat. Acad. Sci., Wash.*, 1958, **44**, 885
[26] Preiss, J., Berg, P., Ofengand, E., Bergmann, F. and Diekmann, M. *Proc. nat. Acad. Sci., Wash.*, 1959, **45**, 319
[27] Schweet, R., Bevard, F., Allen, E. and Glassmann, E. *Proc. nat. Acad. Sci., Wash.*, 1958, **44**, 173
[28] Stephenson, M. L. and Zamecnik, P. C., *Proc. nat. Acad. Sci., Wash.*, 1961, **47**, 1627
[29] Hoagland, M. B., Stephenson, M. L., Scott, J. F., Hecht, L. I. and Zamecnik, P. C. *J. biol. Chem.*, 1958, **231**, 241
[30] Zamecnik, P. C., Stephenson, M. L. and Hecht, L. I. *Proc. nat. Acad. Sci., Wash.*, 1958, **44**, 73
[31] Crick, F. H. C. *Symp. Soc. exp. Biol.*, 1958, **12**, 138
[32] Lagerkvist, U. and Berg, P. *J. mol. Biol.*, 1962, **5**, 139; Berg, P., Lagerkvist, U., and Dieckmann, M. *J. mol. Biol.*, 1962, **5**, 159
[33] Littlefield, J. W., Keller, E. B., Gross, J. and Zamecnik, P. C. *J. biol. Chem.*, 1955, **217**, 111
[34] Schweet, R. and Bishop, J. In *Molecular Genetics*, Pt. 1, (Ed. Taylor), p. 353. New York; Academic Press, 1963
[35] Bloemendal, H., Littauer, U. Z. and Daniel, V. *Biochim. biophys. Acta*, 1961, **51**, 66; Hoagland, M. B. and Comly, L. T. *Proc. nat. Acad. Sci., Wash.*, 1960, **46**, 1554
[36] Davern, C. I. and Meselson, M. *J. mol. Biol.*, 1960, **2**, 153
[37] Riley, M., Pardee, A. B., Jacob, F. and Monod, J. *J. mol. Biol.*, 1960, **2**, 216
[38] Naono, S. and Gros, F. *C.R. Acad. Sci. Paris*, 1960, **250**, 3889
[39] Jacob, F. and Monod, J. *J. mol. Biol.*, 1961, **3**, 318
[40] Hershey, A. D. *J. gen. Physiol.*, 1953, **37**, 1
[41] Volkin, E. and Astrachan, L. *Virology*, 1956, **2**, 149; *Biochim. biophys. Acta*, 1958, **29**, 544
[42] Normura, M., Hall, B. D. and Spiegelman, S. *J. mol. Biol.*, 1960, **2**, 306
[43] Brenner, S., Jacob, F. and Meselson, M. *Nature, Lond.*, 1961, **190**, 576
[44] Gros, F., Gilbert, W., Hiatt, H. H., Attardi, G., Spahr, P. F. and Watson, J. D. *Nature, Lond.*, 1961, **190**, 581
[45] Hall, B. D., and Spiegelman, S. *Proc. nat. Acad. Sci., Wash.*, 1961, **47**, 137
[46] Yoshida, A. and Tobita, T. *Biochim. biophys. Acta*, 1960, **37**, 513
[47] Dintzis, H. M., *Proc. nat. Acad. Sci., Wash.*, 1961, **47**, 247

CHAPTER 12

THE GENETIC CODE

Each DNA molecule, in most preparations, contains 10^4 complementary nucleotide pairs and, in the large bacteriophage DNA, nearer 10^5 such pairs. Thus within any one molecule of DNA there are at least $4^{10,000}$ that is $10^{6,000}$, possible sequences of nucleotides along each of the strands. Thus a basically identical structure for all DNA molecules can vary in an enormous number of ways with respect to base sequences.

With singly stranded ribosomal RNA of molecular weight of the order of 10^6, the possibilities of variation are about the same and the base sequence appears to be the only structural basis for specificity in both types of nucleic acid. When this implication of the structure is taken in conjunction with the genetic function of both nucleic acids, with the linearity of hereditary determinants within the functional gene, and with the structural linearity of the amino acids in the protein molecules (enzymes) which must be presumed to be the first manifestation of the action of genes, then it seems a natural hypothesis that the base sequence of a nucleic acid determines the sequence of amino acids in a particular protein. This sequence hypothesis has been widely held for nearly 10 years, though only during 1961 and 1962 can it be said that direct experimental evidence was obtained about the relationship between the sequence of bases and amino acids.

About 20 different amino acids occur in all proteins, whether derived from animals, plants or micro-organisms. These 20 are, in alphabetical order, with their short notation in parentheses: alanine (ala), arginine (arg), asparagine (aspN), aspartic acid (asp), cysteine (cys), glutamic acid (glu), glutamine (gluN), glycine (gly), histidine (his), isoleucine (ileu), leucine (leu), lysine (lys), methionine (met), phenyl-alanine (phe), proline (pro), serine (ser), threonine (thr), tryptophan (try), tyrosine (tyr), valine (val). A few other amino acids occur only in particular proteins (for example, hydroxyproline in collagen) and cystine may be regarded as arising from two cysteine molecules. The 20 amino acids listed are of universal occurrence.

The validity of the postulated connexion between amino acid sequence and gene action has been confirmed as more and more such sequences in proteins have been determined by hydrolytic and chromatographic methods. The first demonstration of this was the proof[1] that certain abnormal human haemoglobins, A, S and C, differed from each other with respect to only one amino acid residue in one of the polypeptide chains (Chapter 10).

Since this discovery, other mutant human haemoglobins have been examined and other amino acid substitutions have been found as corollaries, not all at the same location as A, S and C (summarized by Smith[2]). Homologous proteins from various species often differ by only a few, or even one, amino acid. Thus the following amino acids have been found to be interchangeable[2-4]:

Cytochrome c: ala, glu, leu; ala, ser, glu, leu; val, phe; thr, ser
Insulin: ala, thr; val, ileu
γ-globulins: phe, tyr
Tryptophan synthetase A: gly, glu, arg; glu, val; glu, ala; arg, ser
Hormones of the neurohypophysis: ileu, phe; arg, leu

The foregoing refer to naturally occurring mutants of one organism (for example, human haemoglobin and tryptophan synthetase) or to natural differences between proteins of different species since there are clear genetic differences in these cases. (The induction of such replacements of amino acids through chemical action on nucleic acids will be discussed later.) Since these genetic differences controlling amino acid sequence can only be inherited through DNA, or RNA in the case of some viruses, the relation of this variation to that of the nucleic acids must be examined.

The variable components of both types of nucleic acid are the four bases characteristic of their component nucleotides, namely adenine (A), guanine (G), cytosine (C) and thymine (T) in DNA and A, G, C and uracil (U) in RNA (capital letters of the bases are used to denote the corresponding component nucleotide in the nucleic acid as well as, in the relevant context, the actual proportion of these nucleotides present). The question naturally arises 'How can the sequence of more than 20 different amino acids be guided by only 4 kinds of base?' This is called the coding problem of the transfer from nucleic acid to protein of genetic 'information', the specification of the amino acid sequence in a given protein. This specification is probably adequate also to determine the spatial configuration of the protein, judging from the precise relation of the determined shape of myoglobin to its amino acid sequence. This transfer of genetic information is almost certainly mediated through RNA, in various forms (*see* Chapter 11), so that the coding problem has at least 3 aspects:

I. The translation of the genetic information in the RNA into the amino acid sequences of a particular protein.

II. The transference of the genetic information in the nucleotide sequence of DNA to some form of RNA.

III. The molecular mechanism of variation within the DNA.

THE PROTEIN–NUCLEIC ACID CODE

There was initially much study of the principles of any translational code whereby a sequence of 20 amino acids could be determined by 4 bases in a nucleic acid, whether this was regarded as DNA or an intermediary RNA. As early as 1950, before the double-helical structure of DNA had been elucidated, Caldwell and Hinshelwood[5] suggested that the 'code' for each amino acid might consist of two adjacent nucleotide units. In 1954, when the helical structure of DNA had been suggested, Gamow[6] proposed a code according to which three nucleotide bases determined one amino acid, adjacent triplets of bases overlapped and more than one triplet of bases coded for a particular amino acid (the code was 'degenerate'). This overlapping, degenerate, triplet code, and a corresponding non-overlapping triplet code, may be represented as follows, where each horizontal set of three letters represents the code for an amino acid (see also *Figure 8.1*).

Imaginary sequences of
four bases: U G A G T T A U A A T G

An overlapping triplet code: U G A

Coding ratio
$= \frac{1}{3} + \frac{1}{3} + \frac{1}{3} = 1$
         G A G
            A G T
               G T T etc.

A partially overlapping
triplet code: U G A

Coding ratio
$= \frac{1}{2} + 1 + \frac{1}{2} = 2$
          A G T
             T T A
               A U A etc.

Non-overlapping code: U G A

Coding ratio = 3
            G T T
               A U A
                  A T G etc.

The advantage of this code was that it explained very directly how 4 bases could be a code for 20 amino acids, since there are 20 possible combinations of bases, *disregarding order*, when these are chosen in sets of 3 from the 4 available.

The code was an overlapping one and this imposed severe restrictions on the frequency with which any one amino acid should follow another in the proteins of any given species and, indeed, in all living organisms, if the code is universally the same. However, analysis of the known sequences of even a few proteins was sufficient to disprove Gamow's particular proposal that there were restrictions of this kind: indeed Brenner[7] was able to show that if the code were universal the known amino acid sequences rendered unlikely in principle all the various overlapping triplet codes (degenerate or not) that might be proposed.

Even on the assumption that 3 bases are the code for a particular amino acid, the different triplet codes vary in their *coding ratio*, which may be defined as the ratio of the number of bases in a nucleic acid to the number of amino acids in the sequence for which these bases are the code, when the effects at the ends of the molecule are ignored. The coding ratio is thus an important quantity which must be determined for different species of living organisms. For non-overlapping codes, which will alone be considered from now on, this coding ratio is of course, simply the number of bases determining each individual amino acid.

The codes which have been discussed so far are all triplet codes since a doublet code, a pair of bases in sequence determining a given amino acid, would only allow $16 (= 4 \times 4)$ permutations, which is not enough to code for 20 amino acids. But the converse problem with any triplet (or higher) code is that too many $(64 = 4 \times 4 \times 4)$ different combinations of nucleotides are possible. This difficulty of any triplet, or quadruplet or higher, code might be met if (*i*) only 20 of the triplets code for an amino acid, that is, make 'sense' and the remaining 44 triplets are 'nonsense'; (*ii*) the code is degenerate, more than one triplet coding for a given amino acid.

These solutions have to be considered in conjunction with another problem. How may the code be read? That is, how can the correct triplets (or quadruplets, etc.) be selected from a continuous sequence of bases? Various suggestions have been made: (*a*), after every triplet (or quadruplet, etc.) there is a 'comma', a base or 'group' of bases, which does not take part in the coding but separates two coding triplets; (*b*), certain triplets make sense and others make nonsense in such a way that any sequence of bases can be 'read' in only one way, even if no starting point is given (as would most lines on this page if all spaces and punctuation were to be removed); or (*c*), the correct choice is made by starting at a fixed point and working along the sequence of bases three (or four, etc.) at a time.

Various 'comma-less codes' fulfilling the requirements of (*b*) were devised. That of Crick, Griffith and Orgel[8] satisfied the numerical requirement of (*i*), which implies that some groupings are nonsense, Twenty triplet combinations can be chosen out of the 64 positions so as to make sense and so that any sequence of them can be read to make sense in only one way. An example of such a code of 20 triplets is:

$$
\begin{array}{ccccc}
 & & & & \text{A} \\
 & \text{A} & \text{A} & \text{A} & \text{A} & \text{U} \\
\text{A U} & & \text{G U} & & \text{U T G} \\
 & \text{U} & \text{U} & \text{G} & \text{G} & \text{T} \\
2 & + & 6 & + & 12 & = 20
\end{array}
$$

The example given above to illustrate overlapping was constructed using this code. No sequence of the triplets that code will give one of the allowed triplets in a 'false' position, that is overlapping another one. This code is now chiefly of interest as illustrating this theoretical phase of the attack in the coding problem. Experimental information relevant to the problem, other than a few amino acid sequences, was at first hard to obtain but the theoretical approach proved its value in enabling the right questions to be asked. The following 6 questions may now be formulated.

(1) Is the code overlapping? A negative answer seems to be demanded by sequence analysis of proteins[7] but more direct evidence is desirable.

(2) What is the coding ratio? If the code is not overlapping this ratio will be the number of bases which code for each individual amino acid.

(3) Is the code degenerate?

(4) Does only part of the DNA, or RNA, code an amino acid sequence? That is, is part of the nucleic acid sense and part nonsense? Proposals (*i*), (*a*), and (*b*) above require this.

(5) Is the code universal? The translation from base sequence to amino acid sequence may not be the same in all organisms.

(6) How may the code be read unambiguously? The possibilities seem to be (*a*), (*b*) and (*c*), above.

Experimental Evidence Bearing upon the Coding Problem
The composition of nucleic acids and proteins in different organisms

One of the major difficulties which the coding hypothesis had to face became apparent when it was shown that although the DNA in different micro-organisms varied widely in composition (with $(G+C)/(A+T)$ varying from 0·45 to 2·7), the RNA showed much less variation ($(A+U)/(G+C)$ from 1·03 to 1·45) and the protein composition even less[9]. There appeared to be only a weak correlation between the composition of DNA and that of RNA. The amino acid composition of the bulk protein of 11 different bacterial species have been compared with the DNA composition of the same species[10]. Twelve out of 18 amino acids increased, or decreased steadily, though slowly, with GC content of the DNA; 6 showed no correlation. This suggests that the amounts of nonsense DNA cannot be as large as had to be proposed to explain the variation in DNA composition. The existence of correlations indicates that the same code operates in these 11 bacteria (and also in *Tetrahymena pyriformis*). The variation in DNA composition is particularly significant in view of the small heterogeneity of composition of the DNA in micro-organisms, in comparison with that of higher organisms. If the mean GC contents of DNA of two bacterial species differ by more than 10 per cent, there are few DNA molecules of the same GC content in common to the two species. Yet the GC content of bacterial DNA varies from 25 to 75 per cent[10]. Explanations for these observations which have been advanced include the following: (*a*) only a minor part (*ca.* 10 per cent) of the DNA codes protein; the rest is nonsense; (*b*) the DNA-to-RNA translation mechanism varies; (*c*) the code is degenerate; (*d*) the code is not universal; (*e*) the nucleic acid code has less than 4 letters; (*f*) the amino acid composition of proteins performing the same function varies. Of these (*a*), (*c*) and (*d*) are possible and (*b*), (*e*) and (*f*) are less likely on a variety of grounds.

Protein synthesis under the control of synthetic polyribonucleotides

Matthaei and Nirenberg reported in 1961 the isolation from *E. coli* of a cell-free system, consisting of ribosomes, a supernatant from centrifugation at 10^5 g, adenosine triphosphate, a system generating this substance and buffers[11]. This system synthesized proteins when polyribonucleotides were added, whether naturally occurring RNA or synthetic polyribonucleotides. Strikingly, they found that the addition of singly-stranded polyuridylic acid (poly-U: —U—U—U—U—U— etc.) caused the production by this system of poly-phenylalanine, which was isolated from the products insoluble in trichloracetic acid. Poly-U in a 2-stranded or 3-stranded complex with poly-A had no activity. The longer poly-U chains were more active than shorter ones and one molecule of poly-U directed the synthesis of a number of molecules of poly-phenylalanine. Thus each coding unit functions catalytically for a limited time[11]. An amino acid intermediate in which phenylalanine is linked to soluble RNA has been detected. This approach was extended by this group[11] and by another at New York[12], in the following way. Synthetic polyribonucleotides were prepared by means of the polynucleotide phosphorylase already described (Chapter 5): these contain one, two or three different kinds of ribonucleotides. The results with poly C, poly A and poly G were not clear-cut. But the synthetic copolymers, which were single chains and were presumably randomly ordered with

TABLE 12.1

Nucleotide Code Assignments for Amino Acid Incorporation
Stimulated by Polyribonucleotides in a Cell-free System from *E. coli*

Amino acid	Code*	Amino acid	Code*
ala	UCG	leu	UUC (UUG, UUA)‡
arg	UCG	lys	UAA
aspN	UAA‡ (UAC)‡	met	UGA
asp	UAG‡	phe	UUU
cys	UUG or UGG§	pro	UCC
glu	UGA	ser	UUC+UCG§
gluN	UCG†	thr	UAC‡+UCC‡
gly	UGG	try	UGG
his	UAC‡	tyr	UUA
ileu	UUA	val	UUG

* Written without regard to order
† Predicted from replacement data, not from the incorporation experiments.
‡ Reference 12 only.   § Reference 11 only.
Triplet codes with no symbols represent those found by both groups[11, 12].

respect to base sequence, directed the incorporation of 14 amino acids into acid-precipitable protein in a highly specific manner.

The results were analysed by assuming a triplet code so that the frequency of random occurrence of any given triplet in a particular polyribonucleotide copolymer relative to the triplet UUU could be calculated from its composition. This frequency was then compared with the incorporation of individual amino acids relative to phenylalanine, for which poly-U is the specific activator. The triplet codes, often called 'codons', obtained by the two groups of investigators agree remarkably well and are given in Table 12.1, without regard to the order of bases within each triplet, since this cannot at present be determined. With 11 of the amino acids the agreement between relative incorporation and the calculated ratio of triplet to UUU was good: the discrepancies of the others could be the result of slight deviations from randomness in the base sequences of the copolymer.

The contribution of these studies to answering questions 1–6 may be summarized as follows.

(1) and (2) Since the coding units of alanine, arginine, aspartic acid, glutamic acid, histidine, methionine, possibly threonine, and glutamine (by a different argument) all contain three different nucleotides, the minimum coding ratio appears to be three, which would agree with a non-overlapping triplet code.

(3) Three amino acids, asparagine, leucine and threonine, were coded by more than one triplet. The code therefore appears to be degenerate for these amino acids, and possibly also for cystine and serine, though the codes given in Table 12.1 may represent only a temporary discrepancy in the observations. Thus the code may be said to be at least partially degenerate in this type of experiment. The number of triplets of the same composition actually assigned to amino acids never exceeded half the possible number of permutations. So it is possible, though not proven, that no complement of a triplet is itself a triplet (complement in the sense of pairing A with T and G with C).

(4) Poly-A and poly-AG do not direct incorporation of amino acids so that nonsense base sequences appear to exist, and this would preclude the possibility of a *completely* degenerate code, as distinct from the partially degenerate one which the results seem to require. Altogether 24 triplets have been listed as coding amino acids, 15 containing one U, eight containing UU, and one containing UUU. But of the 64 possible triplets 37 contain at least one U. There are thus 13 triplets containing U and 27 not containing U which have not yet been assigned in the code. How many of these are nonsense cannot be decided by this kind of experiment, which is limited in its sensitivity to triplet codes containing U. Only peptides insoluble in trichloracetic acid are studied in this system and peptides containing more than about five phenylalanine residues, coded by UUU, are insoluble. This might account for the dominance of uracil in these codes—a feature which leads to absurd values for the composition of RNA deduced from the overall protein composition of known organisms[13]. In any case, the cell-free incorporation process is much less selective than actual protein synthesis and also different in other respects.

(5) As will be discussed in more detail below, the coding units functioning in this *E. coli* system are similar to those directing protein synthesis by TMV and its mutants and in accord with the observed

amino acid replacements in human haemoglobins; so that part at least of the code is universal.

Chemical induction of mutants with altered protein

When the RNA of TMV is treated with nitrous acid, nucleotides are deaminated and mutant strains of the virus are produced. The chemical changes that occur are:

$$C \rightarrow U; A \rightarrow \underset{\text{cell}}{\overset{\text{in}}{\text{Hypoxanthine}}} \rightarrow G; G \rightarrow \text{xanthine (not mutagenic)}.$$

The mutant viruses produce strains which, in many cases, differ from that produced by wild-type virus in that a single amino acid has been replaced by another in the protein of the virus. Thus one nitrous acid mutant exhibited the change: leucine → phenylalanine. The corresponding code change would be, from Table 12.1, UUC → UUU, that is

TABLE 12.2

Amino Acid Replacements in HNO_2 Mutants of Tobacco Mosaic Virus
(after Speyer and colleagues[12])

Replacement	Times observed	Code change	Agreement†
asp* → ser	4	UAG, UAA (UAC) → UUC	−
asp* → ala	6	UAC → UGC	+
asp → gly	2	UAG → UGC	+
arg → gly	5	UCG → UGG	−
glu* → gly	1	UAG → UGG	+
glu* → gly	2	UAG → UGG	+
gluN → val	1	(UCG)‡ → UUG	
glu* → val	2	(UCG) → UUG	
ileu → val	1	UUA → UUG	+
leu → phe	1	UUC → UUU	+
pro → leu	2	UCC → UUC	+
pro → ser	3	UCC → UUC	+
ser → leu	1	UUC → UUC (UUA,UUG)	−
ser → phe	3	UCC → UUC	+
thr → ser	1	UCC → UUC	+
thr → ileu	7	UAC → UAU	+
thr → met	3	UAC (UCC) → UAG	−
tyr → phe	1	UUA → UUU	−

* asp* may be aspartic acid or asparagine; glu* may be glutamic acid or glutamine.
† +Reported replacement agrees with code letter assignment for *E. coli* system, assuming C → U and A → G through HNO_2 treatment.
  − Lack of agreement on this basis
‡ Predicted letter for glu N used.

Observations collected in the Table are those of Tsugita and Fraenkel-Conrat (*J. mol. Biol.*, 1962, **4**, 73), Tsugita (*Protein, Nucleic Acid, Enzyme* (*Tokyo*), 1961, **6**, 385) and Wittmann (*Naturwissenschaften*, 1961, **48**, 729).

the replacement of C by U, which is in accord with the known chemical action of nitrous acid. Table 12.2 summarizes the extent of the agreements of this kind[12]. The agreement of two-thirds of the known replacements is excellent considering that there is always a chance that some of the less frequently observed virus mutants are spontaneous and considering the very different nature of the two systems. It was this agreement which is referred to in Table 12.2 and under (5), page 147. There are, however, certain discrepancies (−in Table 12.2) and, in these instances, the agreement may be incomplete because of undetected U-codes or of additional non-U letters.

In an overlapping triplet code, alteration of one base should change two or three adjacent amino acids in the polypeptide chain, according to the degree of overlap. In a non-overlapping code only one amino acid should be altered and this is what is usually observed as a result of treating TMV RNA with nitrous acid. In the rare cases when two amino acids are altered, by the occurrence of two separate deaminations in the RNA, changes are not at adjacent positions in the polypeptide chain.

In a similar type of experiment, mutations were induced in *E. coli* by ultra-violet light, x-rays or by ethyl methane sulphonate and an enzyme protein, alkaline phosphatase, was isolated and digested to examine changes in peptide sequences[14]. The mutants were also examined genetically. The substitution of valine for alanine has been observed as the result of the mutagenic action of ethyl malonate sulphate which is considered to change a GC base pair to AT. Close linkage of genetic sites in two other mutants was also shown to be paralleled by close linkage of the amino acids controlled by the sites. Similar studies have been initiated on the genetically controlled structure of the lysozyme produced in bacteria infected by normal and mutant T4 bacteriophage[15].

Amino acid replacements in proteins

The differences between mutant forms of the same protein from a given species (for example, human haemoglobin) or between homologous proteins from different species (for example, cytochrome *c*, insulin) often amount to only one amino acid change amongst hundreds present in the protein. These changes, some of which have been already mentioned, have been examined by E. L. Smith in relation to the *E. coli* code for 14 amino acids[2]. Smith made the restrictive assumption that each mutation involves replacement of only one base in each triplet and he was able to predict certain codes unknown at the time. Thus, knowing that the codes for lysine, valine

148

and glycine are UAA, UUG and UGG respectively, and that the replacements glu→lys, val or gly occur in human haemoglobins (*see* above, page 121), it was deduced from the first replacement that the code for glutamic acid must contain at least one A, from the second at least one U, and from the third at least one G. Thus the code for glutamic acid was predicted to be UGA and this was later independently confirmed with the *E. coli* system by Speyer and colleagues[12]. Similar predictions were made for alanine (UCG), asparagine (UAC) and aspartic acid (UAG) on the basis of replacement date for cytochrome *c*, and insulin as well as for human haemoglobin. The correctness of these predictions is strong evidence for the universality of the genetic code. It was further concluded that in 16 of the codes containing at least one U (out of 37 theoretically possible) the same position is occupied by U.

In the studies just described, knowledge of amino acid replacements in proteins was used to check and extend the amino acid triplet code deduced from experiments on control of protein synthesis by polyribonucleotides. However, persistent attempts have been made, ever since the earlier proposals of Gamow, to determine a universal triplet code from the known amino acid replacements[16–18]. The main principle employed by Zubay and Quastler[18] was that the number of nucleotide changes associated with the replacement of a single amino acid with unchanged neighbours should be minimized. Extensive recent data on amino acid replacements were employed (in their

TABLE 12.3
Triplet Codes

Amino acid	Woese[17]	Zubay and Quastler[18]	E. coli system (Table 9.1)
ala	UAG	UCG	UCG
asp*	GAU	UCA	UAG asp
			UAA aspN
glu*	UAU	UUA	UGA glu
			UCG gluN
gly	GAG	UUG	UGG
leu	UCG	UUC	UUC
lys	CCG	UGA	UAA
phe	UUG	UUU	UUU
pro	CCC	UCC	UCC
ser	AAG	UGG	UUC+UCG
thr	CAC	UAG	UAC+UCC

* The acid and its amide.

149

paper Zubay and Quastler[18] give a useful compilation) and a code was proposed which is given in Table 12.3, together with that of Woese[17] and the one deduced from the *E. coli* system (Table 12.1). Only the deductions considered by Zubay and Quastler to be their most probable are included. There is thus some agreement now between these two approaches although the analysis of the replacement data still seems to be possible only if certain of the incorporation data deduced from the *E. coli* system are also used in the argument.

Genetic studies on acridine mutants of bacteriophage

It has been shown that mutations can be induced in the r_{II} region of the *E. coli* bacteriophage T4 by various substances, including base analogues (for example, 5-bromouracil), nitrous acid and various acridines, including proflavine (3,6-diamino acridine), 9-amino acridine, and acridine orange (3,6-bisdimethylaminoacridine).

The mutants appeared to be of two mutually exclusive classes[19, 20] according to their susceptibility to reversion by 5-bromouracil or by acridines: the group revertible by acridines included mutations themselves induced by acridines, such as proflavine. On these grounds, it was suggested that acridines act as mutagens by adding or deleting a base or bases in the nucleic acid. The most important evidence for this was that mutants produced by acridines are seldom 'leaky' since they are almost always completely lacking in the function of the gene. If an acridine mutant is produced by adding a base, then it should revert to wild type by deleting one.

This system has been exploited by Crick and colleagues[21] who, in a genetic analysis of the reversional behaviour of acridine mutants, were able to determine some of the fundamental characteristics of the genetic code. The r_{II} region of T4 bacteriophage consists of two adjacent genes or 'cistrons'* (A and B). The wild type will grow on two particular strains of *E. coli* (namely *B* and *K12* (λ), *K*, for short); if the function of either of the genes is lost it will not grow on *K* and forms an *r* plaque on *B*. It was a mutation of this kind (in the *B* cistron) induced by proflavine, which was studied. (In this system, a 'leaky' mutant is one which, although growing on strain *K*, does not show the wild-type plaque on strain *B*.)

The genetic studies showed that this mutation (which will be denoted by +) reverts not by reversing the original mutation at the same point on the genetic map but by producing a second mutation (denoted by −) at a nearby point. This second mutation is a 'suppressor' within the same gene. All of the 18 suppressors of the

* *See* Glossary, Page 185.

original mutation when acting alone had themselves lost the function of the gene in the sense given above. The double mutants (+ and −) grew on K and had a variety of plaque types on B, many of these like the wild type. Pure strains containing a suppressor (−) were isolated genetically and suppressors of these (denoted by + again) were mapped. These new suppressors, when acting alone, had also simply lost the function of the gene. Again, suppressors (−) of this last set of suppressors have been observed and mapped. Altogether about 80 independent mutants were observed: suppressors (−) of the original mutant (+); suppressors (+) of the first suppressors; suppressors (−) of the second set of suppressors (+). The explanation of these observations given by the authors was as follows.

It was assumed that the B cistron controls the production of a polypeptide chain; that a string of nucleotide bases is read, triplet by triplet, *from a fixed starting point* on the left of the B cistron (*Figure 12.1 (a)*); and that the mutation was induced by proflavine by insertion of an additional base in the nucleotide sequence which produces wild-type phage. The reading of all triplets to the right of this point of addition would be shifted along by one base and so be incorrect, thereby producing a protein of altered amino acid sequences from that point onwards (*Figure 12.1 (b)*). This would explain why the function of the gene is lacking. In the first suppressors (−) a base is deleted (*Figure 12.1 (c)*). If this deletion is present alone, the sequence will again be wrong at all points to the right of the original deletion and the gene will be lacking, as with the previous + mutation. However, if both mutations are present in the same piece of DNA (*Figure 12.1 (d, e)*), as in the original suppressed mutant (+ and −), then the reading of triplets will only be 'wrong' over the smaller limited region between the points of the + and − mutations. This would explain why the function of the gene is nearly restored, but rarely quite completely, since a small length of amino acid sequence is different from that of the wild type. On this basis, certain predictions could be made and tested. Thus a combination of type + with + (or − with −) should not give wild type but a mutant lacking the gene (that is, growing on B to give r plaques). This was confirmed. Combinations of + with − will differ in the way shown in (*d* and *e*) of *Figure 12.1*, the reading frame being displaced oppositely. The changes produced in a given sequence will differ according to whether the displacement of this frame is ← or →. Examination of the location of − suppressors relative to + mutants suggested that in certain regions the shift ← led to unacceptable readings of triplets. The combination of a + with a − was phenotypically r (that is, gene function lost) if it involved a ←

shift over these regions. Otherwise a + combined with a − led to the wild (or pseudo-wild) form being restored. Predictions of phenotypes for ± combinations were correct and thereby confirmed the correctness of this interpretation.

The theory was framed in terms of a triplet code, even though the evidence so far described need not be limited to this. Positive evidence

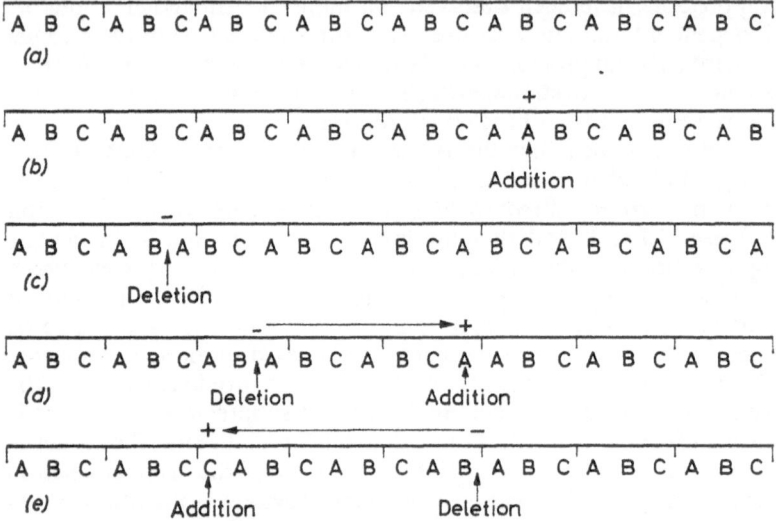

Figure 12.1. Additions and deletions in a coding sequence. A,B,C each represent a different base of the nucleic acid[21]. For simplicity a repeating sequence of bases, A B C, is shown. (This would code for a polypeptide in which every amino acid was the same.) A triplet code is assumed in drawing the lines representing the 'reading frame', which shows how the code is read in sets of three starting from the left. (a) The original sequence; (b) Addition of a base; (c) Deletion of a base; (d) and (e) Addition and deletion together. Arrows show the effective shift in the reading frame caused by additions and/or deletions

(Reproduced by courtesy of the Editor of Nature, London)

for a triplet code came from the observation that triple mutants of the form + + + or − − −, when they did not involve ← shifts across the 'unacceptable' regions were all wild or pseudo-wild. This was exactly what was expected in favourable cases, if the coding ratios were 3 or a multiple of 3. The possibility of finding the coding ratio in this way must depend on the gene-product functioning with little disturbance when one amino acid has been added to or deleted from it. Taken in conjunction with other evidence on acridine mutants, it was concluded that a coding ratio of 3 was more probable than a multiple of 3.

It was also concluded that the code probably did not contain nonsense since if only 20 out of the 64 possible triplets were sense, the region of the genetic map containing − suppressors of the original + mutant should be much smaller than was observed. Thus the triplet code is probably degenerate. Another interesting conclusion from this genetic study is that the amino acid sequence of the pseudo-wild protein produced by one of the double mutants (±) should be altered over a range and not just at two points only. This is a prediction which may be tested in cases where a particular protein which is subject to known genetic control may be isolated and analysed, for example, the lysozyme of phage or alkaline phosphatase of *E. coli*.

Determination of nucleotide sequences in nucleic acids

Ultimately the problem of the genetic code can only be regarded as solved when the amino acid sequence of a protein can be read off from the nucleotide sequence in the particular nucleic acid which controls its formation. Progress that has already been obtained in determining nucleotide sequences has been described in Chapters 4 and 5. A start has been made in relating such sequences to particular amino acids by the work of Lagerkvist, Berg and Dieckmann[22] on soluble (transfer) RNA. They determined 11 different sequences, between 4 and 9 nucleotides long, next to the common pCpCpA chain end of this RNA (p refers to a phosphoric acid residue; when written on the left of a letter denoting a base it signifies the 5′ phospho-ester; on the right it signifies the 3′ ester). The chemical heterogeneity of such RNA was thereby confirmed.

The sequences adjacent to the end triplet were not the same for the S-RNA which transferred isoleucine and leucine. The S-RNA which combined with isoleucine terminated in the sequence−−GpCp (UpC)pApCpCpA. The S-RNA chains with leucine were of two separable types, namely−−GpCpApCpCpA and−−GpUpApCpCpA. This indicates not only that there is a direct correlation between nucleotide sequence and a particular amino acid but also that more than one nucleotide sequence codes for leucine. The genetic code for leucine must therefore be degenerate. This conclusion has been confirmed for leucine by the separation[23] of leucine-transferring soluble RNA into two fractions which are involved in leucine incorporation in a cell-free ribosomal system under the stimulus of two different polyribonucleotides.

Transference of genetic information between different species

If the ribosomes of one species were to react with the soluble (transfer) RNA of another and thereby to incorporate amino acids into the protein made by the ribosomes, then this would support the universality of the genetic code. Such interchange of genetic information has been observed between mammalian and bacterial species. Thus haemoglobin is synthesized in mammalian red cell reticulocytes from amino acids transferred by S-RNA from *E. coli* and other micro-organisms[24]. This constitutes good evidence for the universality of the genetic code.

The Nature of the Genetic Code

The foregoing evidence, mostly published in 1961 and 1962, leads to the following answers to the questions already posed (page 143).

(1) The code is not overlapping.

(2) The coding ratio is three and, because of conclusion (1) a triplet of nucleotides is the code for each amino acid.

(3) The code is at least partially degenerate.

(4) Some 'nonsense' nucleic acid may exist but not a large proportion of the whole.

(5) The code is probably universal.

(6) The code is read by starting at a fixed point and working along the sequence of bases. The nature of this 'fixed point' is unknown.

THE TRANSFERENCE OF INFORMATION FROM DNA TO RNA

It had been assumed that the genetic information stored in DNA was transmitted to proteins through the mediation of RNA, and the discovery of the existence of 'messenger RNA' as a short-lived intermediate product of the structural genes has now confirmed this presumption. The evidence and detailed role of this RNA, and of soluble and ribosomal RNA have been discussed more fully in Chapter 11. The present concern is with the molecular mechanism by which the coding information in the nucleotide sequences of DNA might be transferred to RNA. Experimental information is sparse and may be summarized thus.

(*a*) When T2 bacteriophage infects *E. coli*, a small RNA fraction appears characterized by an exceptionally high rate of synthesis[25]. This RNA fraction had the same base ratios as the T2 DNA and on heating and cooling it in the presence of denatured single chain T2 DNA, hybrids of RNA and DNA were formed while no such hybrids were obtained when non-homologous DNA was used[26]. That this 'T2 specific RNA' represents a complementary transcript of T2 DNA was further suggested by evidence from equilibrium centrifugation for the occurrence of natural complexes between DNA and RNA in *E. coli* infected with T2 bacteriophage[27]. (*b*) A nuclear RNA fraction with a very high rate of synthesis has been isolated from calf thymus glands. It has the same base composition as the thymus DNA and appears to be the equivalent in the mammalian cell of bacterial messenger RNA[28]. (*c*) A yeast RNA fraction has been isolated which is characterized by a high rate of turnover and by a base composition similar to yeast DNA[29]. (*d*) It has been shown that RNA is synthesized on a DNA template by a polymerizing enzyme in liver nuclei[30] and in *E. coli*[31]. In the *E. coli* system, the ribonucleotides are incorporated into the polymer in precise proportion to the relative frequencies of the deoxyribonucleotides in the DNA primer (RNA will not act as a primer). This enzyme thus appears to be the transcribing system.

There have been various speculations concerning the coding mechanism in this interaction of DNA and RNA. For example, Zubay[32, 33] thought that DNA in its double-helical form transcribed a specific nucleotide sequence to single RNA wrapped around it, but this mechanism is almost certainly incorrect.

Because of the evidence[27, 34] that RNA can form a specific complex with the DNA from which it had been made, when this is in the denatured form, some authors have suggested that messenger RNA is synthesized on uncoiled DNA through a double-helical complex. In this complex, one chain would be RNA and the other DNA paired complementarily as in DNA itself[35]. Evidence for the existence of double-helical hybrids of polyriboadenylic acid with oligonucleotides of deoxythymidilic acid[36] and with polydeoxyguanylic acid[37] has been obtained. In the latter instance, the hybrid was thermally more stable than the corresponding double helix of chains of polydeoxyguanylic and polydeoxycytidylic acids. Moreover, the hybrid was not attacked either by ribonuclease or by deoxyribonuclease, separately or together. This could be an important feature of the mechanism for transferring the genetic code, for it would give protection against the action of intracellular nucleases. It was also shown[37] that DNA from *E. coli* interacts with the RNA from *E. coli*, so as to alter its density in caesium chloride, but not with other RNA. However, the role of such hybrids *in vivo* cannot be regarded as proven, if only because of the inability of single chain DNA to act as an effective primer in the synthesis of messenger RNA[38].

THE MOLECULAR MECHANISM OF VARIATION WITHIN DNA

If the molecular basis of mutation is alteration in nucleotide sequences in DNA, the problem arises of how bases without a complementary base in the other chain ('unmatched' bases) can be accommodated within the double-helical framework of the DNA. Fresco and Alberts[39] examined the interaction between the polyribonucleotides poly-A or poly-U and the copolymer poly-AU, which contains various proportions of adenine and uracil residues disposed randomly along the chain. The interaction occurred in such a manner that the two-stranded helical complexes formed by hydrogen bonding between adenine and uracil (equivalent to thymine for this purpose) always contained equal numbers of bases from the poly-A, or poly-U, and of complementary nucleotides (U, or A, respectively) in the copolymer, poly-AU. This indicated that the unmatched bases formed loops outside the helical framework and models showed that this was quite feasible. Apparently, the number of mis-matchings which could be tolerated in the DNA helix is quite large and could explain the deviations from the normal analytical relations which have occasionally been reported (*see* page 28).

Fresco and Alberts think that their observations suggest three types of molecular counterpart for the 'point' mutations observed genetically. These are shown in *Figure 12.2* which depicts the template chain of DNA on the left and, on the right, the new chain in which various possible forms of 'mistaken' replications are indicated. In (*a*) a nucleotide (X = T,C,G), which does not match with thymine, has been *substituted* for the required adenine in the new chain. In (*b*), a template residue (T) has looped out of the helix and is unmatched in the new chain. A base pair, therefore, is *deleted* from the new chain. In (*c*), unmatched nucleotides (G,C) have been added to the growing chain and are accommodated by forming a loop. Subsequently, new base pairs will be *added* to the progeny DNA. The substitution mechanism (*a*), it is suggested, could produce reversible mutations and the deletion and addition mechanisms (*b*) and (*c*), could account for irreversible mutations.

Acridines, especially proflavine, are unusual in that their mutagenic action is probably the addition or deletion of a base, since mutants induced by them either do not give the original protein controlled by the gene or else give a greatly altered one[40]. The postulate of an addition effect formed the starting point of the account given by Crick and colleagues[21] of their genetic experiments on a proflavine-induced mutant of the r_{II} region of T4 bacteriophage (although that account would have been equally valid if the action of the acridines had been to delete a base, rather than to add one). The original explanation of

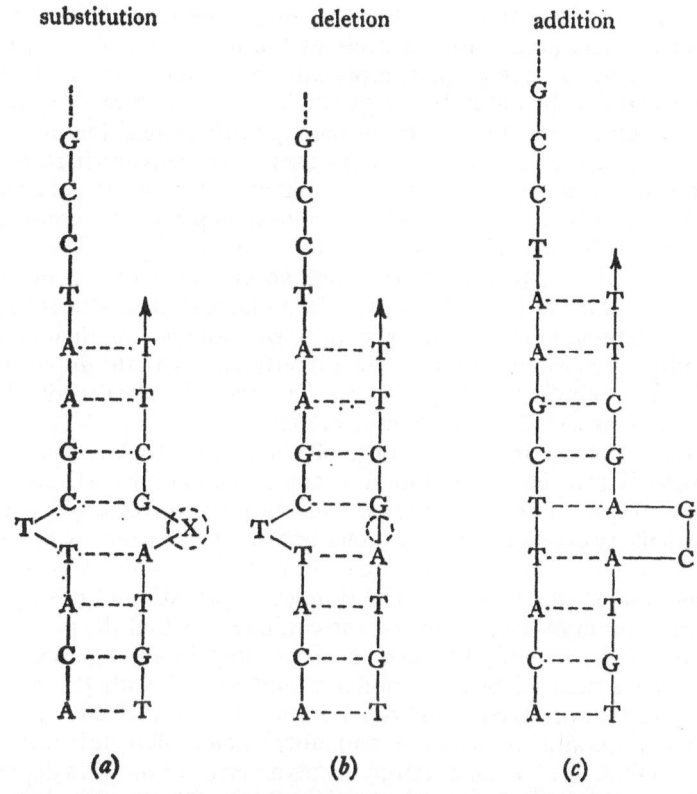

Figure 12.2. Hypothetical models[39] for point mutations G, C, A, T: bases. In each case the left column represents the template and the right column represents the growing chain. Bases in loops outside the double helix are depicted as offset from the vertical columns. The dotted circles indicate a 'mistake' in replication

(Reproduced by courtesy of the Editor of *Proceedings of the National Academy of Sciences of the United States of America*)

Freese[19] for the two classes of mutants, those induced by base analogues and by acridines, in the r_{II} region of T4 bacteriophage was that in one the interchange was $AT \rightleftharpoons GC$ and in the other $AT \rightleftharpoons CG$, so that one of these changes could not be reversed by a change of the other sort. As this did not explain why the two sets of sites for base analogue and acridine mutants should *all* be different, Brenner and colleagues[40] were led to make the proposal already mentioned.

The strong bacteriostatic action of the aminoacridines has been known for a long time and the structural requirements have been studied[41]. The quantitative aspects of the inhibition of *Aerobacter aerogenes* by aminoacridines, especially proflavine, were studied in detail by Hinshelwood and colleagues[42]. They obtained much evidence that, in this system at least, genetic changes with acquisition of resistance were occurring in each cell of the bacterial population in response to the proflavine and not through natural selection of previously existing resistant mutants. This proposal appeared to contradict current biological ideas of the non-inheritability of 'acquired characteristics'. Meanwhile, evidence accumulated of the unusual character of the biological action of the aminoacridines—the strength of their interaction with nucleic acids *in vitro* and *in vivo*, their ability to inhibit phage multiplication, their interference with the duplication of DNA—culminating in the special features, already described, of their mutagenic effects on T4-bacteriophage.

Studies on the interaction of proflavine with DNA showed that there was strong binding of four molecules for every turn of the double helix and that this was not only electrostatic but involved a high degree of specificity dependent on the intactness of the flat aromatic acridine ring system[43]. There was an increase in viscosity when proflavine was bound on helical DNA[44, 45]. The simplest explanation of this and of sedimentation measurements on the complex was that the proflavine intercalated reversibly between the base rings in an extended two-stranded structure. Such a model would accord with the genetic evidence that proflavine, and other aminoacridines, effectively add a base to the coding sequence. Presumably when proflavine is attached to the DNA, the 'reading frame' makes an error of the kind depicted in *Figure 12.1*, (*b*) for a + mutation. Then either removal of the proflavine could allow reversion to the original nucleotide sequence and so to wild type, or, in the process of replication, a new base-pair could be permanently added by addition of a new nucleotide to one (or both) of the DNA chains at the point where the proflavine is interposed. The mechanism (*c*) of *Figure 12.2* seems to depict such a situation. Intercalation as such is not essential to this picture, so long as the proflavine causes an extension of the double helix, and the physicochemical observations do not rigorously prove intercalation[45]. The process whereby most of the genetically controlled DNA in a cell acquires an additional base-pair(s) through the insertion of proflavine might only be a gradual process with just the characteristics observed by Hinshelwood and colleagues[42] in the training procedure when resistance to proflavine is acquired. The recent genetic studies and knowledge of the interaction of the aminoacridines with DNA open up new possibilities concerning mutagenesis which may render obsolete for bacteriophage and bacteria the terminology devised to apply to larger organisms in which the DNA is protected from all external influences.

Another type of explanation of mutagenesis is that a base in the DNA is changed either into a tautomeric form or the hydrogen ion dissociation of one of its groups is altered so that anomalous base-pairing ensues[46-48]. The latter change can be induced by alkylation and it has been shown that ionized guanine (a form enhanced by alkylation) would pair with thymine, instead of cytosine, through two hydrogen bonds in the DNA double helix. Similarly ionized thymine (or 5-bromouracil) would pair with guanine, instead of adenine. Spontaneous mutations could then be attributed to the statistically small, but finite, chance of ionization during replication, and mutations induced by denaturing agents, such as heat or acid, could be similarly explained[49, 50]. The ability of γ-rays to cause a change in the hydrogen ion dissociation curves, with concomitant denaturation, had previously been suggested[51] as a possible molecular basis for the mutagenic action of such radiation. DNA is certainly far more sensitive to the denaturing action of γ-rays than to any other effect[51] (Chapter 9).

REFERENCES

[1] Ingram, V. M. *Nature, Lond.*, 1957, **180**, 326; *Biochim. biophys. Acta*, 1958, **28**, 539

[2] Smith, E. L. *Proc. nat. Acad. Sci., Wash.*, 1962, **48**, 677

[3] Anfinsen, C. B. *The Molecular Basis of Evolution.* New York; Wiley, 1959

[4] Zubay, G. and Quastler, H. *Proc. nat. Acad. Sci., Wash.*, 1962, **48**, 461

[5] Caldwell, P. C. and Hinshelwood, C. N. *J. chem. Soc.* 1950, 3156

[6] Gamow, G. *Biol. Medd. Kbh.*, 1954, **22**, No. 3

[7] Brenner, S. *Proc. nat. Acad. Sci., Wash.*, 1957, **43**, 687

[8] Crick, F. H. C., Griffith, J. S. and Orgel, L. E. *Proc. nat. Acad. Sci., Wash.*, 1957, **43**, 416

[9] Lee, K. Y., Wahl, R. and Barbu, E. *Ann. Inst. Pasteur*, 1956, **91**, 212; Belozersky, A. N. and Spirin, A. S. *Nature, Lond.*, 1958, **182**, 111

[10] Sueoka, N. *Proc. nat. Acad. Sci., Wash.*, 1961, **47**, 1141; *J. mol. Biol.*, 1961, **3**, 31

[11] Matthaei, J. H. and Nirenberg, M. W. *Proc. nat. Acad. Sci., Wash.*, 1961, **47**, 1580, 1588; Matthaei, J. H., Jones, O. W., Martin, R. G. and Nirenberg, M. W. *Proc. nat. Acad. Sci., Wash.*, 1962, **48**, 666

[12] Speyer, J. F., Lengyel, P., Basilio, C. and Ochoa, S. *Proc. nat. Acad. Sci., Wash.*, 1962, **48**, 441

[13] Chargaff, E. *Nature, Lond.*, 1962, **194**, 86

[14] Garen, A., Levinthal, C. and Rothman, F., *J. Chim. phys.*, 1961, **58**, 1068

[15] Streisinger, G., Mukai, F., Dreyer, W. J., Miller, B. and Harrar, G. *J. Chim. phys.*, 1961, **58**, 1064

[16] Gamow, G., Rich, A. and Ycas, M. *Advances in Biological and Medical Physics*, (Ed. Lawrence and Tobias), Vol. 4. New York; Academic Press, 1956

[17] Woese, C. R. *Biochem. biophys. Res. Commun.*, 1961, **5**, 88

[18] Zubay, G. and Quastler, H. *Proc. nat. Acad. Sci., Wash.*, 1962, **48**, 461

[19] Freese, E. *Proc. nat. Acad. Sci., Wash.*, 1959, **45**, 622

[20] Orgel, A. and Brenner, S. *J. mol. Biol.*, 1961, **3**, 762

[21] Crick, F. H. C., Barnett, L., Brenner, S. and Watts-Tobin, A. J. *Nature, Lond.*, 1961, **192**, 1227

[22] Lagerkvist, U. and Berg, P. *J. mol. Biol.*, 1962, **5**, 139; Berg, P., Lagerkvist, U. and Dieckmann, M. *J. mol. Biol.*, 1962, **5**, 159

[23] Weisblum, B., Benzer, S. and Holley, R. W. *Proc. nat. Acad. Sci., Wash.*, 1962, **48**, 1449

[24] von Ehrenstein, G. and Lipmann, F. *Proc. nat. Acad. Sci., Wash.*, 1961, **47**, 941

[25] Volkin, E. and Astrachan, L. *Biochim. biophys. Acta*, 1958, **29**, 536

[26] Hall, B. D. and Spiegelman, S. *Proc. nat. Acad. Sci., Wash.*, 1961, **47**, 137; Nomura, M., Hall, B. D. and Spiegelman, S. *J. mol. Biol.*, 1960, **2**, 306

[27] Spiegelman, S., Hall, B. D. and Storck, R. *Proc. nat. Acad. Sci., Wash.*, 1961, **47**, 1135

[28] Sibatani, A., de Kloet, S. R., Allfrey, V. G. and Mirsky, A. E. *Proc. nat. Acad. Sci., Wash.*, 1962, **48**, 471

[29] Ycas, M. and Vincent, W. S. *Proc. nat. Acad. Sci., Wash.*, 1960, **46**, 804

[30] Weiss, P. *Proc. nat. Acad. Sci., Wash.*, 1960, **46**, 993

[31] Hurwitz, J., Bresler, A. and Diringer, R. *Biochem. biophys. Res. Commun.*, 1960, **3**, 15; Furth, J. J., Hurwitz, J. and Goldmann, M. *Biochem. biophys. Res. Commun.*, 1961, **4**, 362; Ochoa, S., Burma, D. P., Kroger, H. and Weill, J. D. *Proc. nat. Acad. Sci., Wash.*, 1961, **47**, 670

[32] Zubay, G. *Proc. nat. Acad. Sci., Wash.*, 1962, **48**, 456

[33] Zubay, G. *Nature, Lond.*, 1958, **182**, 112

[34] Geiduschek, E. P., Nakamoto, T. and Weiss, S. B. *Proc. nat. Acad. Sci., Wash.*, 1961, **47**, 1405

[35] Rich, A. *Ann. N.Y. Acad. Sci.*, 1959, **81**, 709

[36] Rich, A. *Proc. nat. Acad. Sci., Wash.*, 1960, **46**, 1044

[37] Schildkraut, C. L., Marmur, J., Fresco, J. R. and Doty, P. *J. biol. Chem.*, 1961, **236**, PC2

[38] Burma, D. P., Kroger, H., Ochoa, S., Warner, R. C., and Weill, J. D. *Proc. nat. Acad. Sci., Wash.*, 1961, **47**, 749

[39] Fresco, J. R. and Alberts, B. M. *Proc. nat. Acad. Aci., Wash.*, 1960, **46**, 311

[40] Brenner, S., Barnett, L., Crick, F. H. C. and Orgel, A. *J. mol. Biol.*, 1961, **3**, 121

[41] Albert, A. *The Acridines.* London; Arnold, 1951

[42] Hinshelwood, C. N. *The Chemical Kinetics of the Bacterial Cell.* Oxford; Oxford University Press, 1946; Dean, A. C. R. and Hinshelwood, C. N. *Prog. Biophys.*, 1955, **5**, 1; Dean, A. C. R. and Hinshelwood, C. N.

REFERENCES

Growth, Function and Regulation in Bacterial Cells. Oxford; Oxford University Press, 1966

[43] Peacocke, A. R. and Skerrett, J. N. H. *Trans. Faraday Soc.*, 1956, **52**, 261

[44] Lerman, L. S. *J. mol. Biol.*, 1961, **3**, 18

[45] Drummond, D. S. *B.Sc. Thesis, Oxford*, 1962; Drummond, D. S., Simpson-Gildemeister, F. W. and Peacocke, A. R. *Biopolymers*, 1965, **3**, 135

[46] Watson, J. D. and Crick, F. H. C. *Nature, Lond.*, 1953, **171**, 964

[47] Freese, E. *J. mol. Biol.*, 1959, **1**, 87

[48] Lawley, P. D. and Brookes, P. *Nature, Lond.*, 1961, **192**, 1081; *J. mol. Biol.*, 1962, **4**, 216

[49] Zamenhof, S. and Greer, S. *Nature, Lond.*, 1958, **182**, 611

[50] Freese, E. *Brookhaven Symposia*, 1959, **12**, 63

[51] Peacocke, A. R. and Preston, B. N. *Proc. Roy. Soc.*, 1960, **153B**, 103

PART 4
SYNOPSIS AND ADDENDUM

SYNOPSIS

Part 1: The Background

The importance of nucleic acids in heredity—indirect evidence

Deoxyribonucleic acid (DNA) possesses a number of properties which the genetic material might be expected to show but which do not prove that genes are composed of DNA. Chromosomes are known to contain the genetic material and in general DNA is confined exclusively to the chromosomes. The amount of this nucleic acid increases directly with ploidy. Apart from the period during the division cycle when DNA is being synthesized it appears to be metabolically stable. It also replicates in a manner which preserves the sequence of bases in the molecule. While the base composition of DNA isolated from different organs of the same species is constant, the composition of DNA from different species varies and is characteristic of the species. The fact that many mutagens react directly with nucleic acids has been used as evidence that the genetic material is a nucleic acid (Chapter 2).

The importance of nucleic acids in heredity—direct evidence

Experiments with micro-organisms have shown in a direct way that DNA is the genetic material. If cells of one strain of a bacterial species are incubated in a solution of DNA from another strain of the same organism, some of the cells of the recipient strain acquire properties of the strain from which the DNA is extracted. These newly acquired characteristics are inherited by the progeny of the transformed cells. Studies with bacteriophage labelled with ^{32}P or ^{35}S show that only the nucleic acid component of the phage enters the bacterial cell and therefore that the phage nucleic acid must carry all the genetic information necessary to direct the synthesis of new phage particles by the host cell. In sexual reproduction in *E. coli* it has been shown that the transfer of genes from the donor to the recipient is correlated with the transfer of DNA from one cell to the other. A number of viruses contain ribonucleic acid (RNA) but no DNA. In these organisms RNA carries the genetic information (Chapter 3).

Part 2: The Molecular and Structural Basis

Deoxyribonucleic acid

DNA consists of two helical polynucleotide chains which are intertwined and are cross-linked by hydrogen bonds. The latter link the purine bases, adenine (A) and guanine (G), with the pyrimidine bases, thymine (T) and cytosine (C), respectively. The exigencies of the structure make quite specific pairing of A with T and of G with C. Bases other than A, G, C and T can also be incorporated into the structure. DNA as normally isolated is a mixture of molecules of different composition, with respect to the proportion of bases, and of different molecular weights. The average composition depends on the species from which the DNA is isolated. DNA in bacteriophage and bacteria, and possibly in the chromosomes of higher organisms, is one long double helix and is broken down by shearing during isolation to a random mixture of average molecular weight about $5-8 \times 10^6$. The sequence of bases along the chains in DNA is not random. It is possible to test the complementarity of such sequences by their ability to form 'hybrid' DNA. DNA has been synthesized enzymatically *in vitro* from nucleoside triphosphates when a small amount of intact DNA is used as primer (Chapter 4).

Ribonucleic acid

RNA occurs mainly in the cytoplasm of all cells and is present in particularly large amounts in cells in which rapid protein synthesis is occurring. RNA has the same polynucleotide structure as DNA, with ribose instead of deoxyribose, and uracil (U) instead of thymine. Three main types of RNA have been distinguished at present. *Soluble RNA* (amino-acid transfer RNA, S-RNA or tRNA) which conveys specific amino acids to the ribosomes for protein synthesis, is a single chain of about 80 nucleotides, the sequence of which have been determined in certain cases. Hydrogen bonds are postulated (G bonds to C and A to U). *Macromolecular RNA*, from ribosomes or viruses, has molecular weights in the range 10^5 to 10^7 and is a single polynucleotide chain which folds back on itself in double-helical regions. These form only a part of the molecule, the shape of which is flexible and varied, when not in combination with its appropriate protein. Its average composition differs less from one organism to another than does that of DNA, and $A \neq U$ and $G \neq C$. RNA, as normally isolated, is heterogeneous with respect to molecular weight but homogeneous RNA can be obtained from some plant viruses in which it exists as a single strand. *Messenger RNA (mRNA)* is the carrier of genetic information from DNA to the ribosomes and is a short-lived fraction constituting less than 4 per cent of the total cellular RNA. Its molecular characteristics are not yet well established. It probably has a molecular weight of $1-5 \times 10^5$ and contains various sub-fractions. Polyribonucleotides have been synthesized enzymically from nucleotide diphosphates and have yielded valuable structural information (Chapter 5).

Nucleoproteins

Nucleoproteins are a complex of the negatively charged polynucleotide chain with a positively charged protein, which is relatively rich in basic amino acids; it is the form in which nucleic acids usually occur *in vivo*. In nucleoproteins containing DNA, the latter retains its double-helical structure and acts as a core around which the protein is wrapped or to which it cross-links. In virus and microsomal nucleoproteins containing RNA, the configuration of the RNA is much more dependent on the shape of the protein matrix in which it is embedded. The form of this matrix appears to be determined by the rules of symmetry which govern the packing of sub-units of protein (Chapter 6).

Chromosomes

The structure of chromosomes is dependent on the nature and arrangement of their constituent molecules and ions. Cytochemical, isotopic and enzymic methods together with direct chemical analysis of chromosomes have shown that chromosomes contain DNA, RNA, basic proteins, acidic proteins and divalent cations particularly those of calcium and magnesium. There is considerable evidence from light and electron microscopy that chromosomes are multistranded. This concept is also supported by the type and frequency of structural changes induced in chromosomes by irradiation and by other, less direct, evidence. Contrary and convincing evidence which favours the idea that the chromosome contains only one double helix of DNA has been obtained by autoradiographic studies of *E. coli* nuclear material and chromosome duplication in a number of plants and animals. The study of lampbrush chromosomes in amphibia also lends support to the single-stranded nature of chromosomes. With regard to the linear differentiation of the chromosome the basic consideration is whether the main chromosomal axis is DNA, protein, or both. The evidence from *E. coli* and newt suggests that protein does not form a major part of the axis. In some chromosome models which have been proposed, DNA forms the axis of the chromosome with adjacent DNA molecules joined by means of linking units of low molecular weight (Chapter 7).

Part 3: The Structure of the Nucleic Acids in Relation to their Biological Function

Replication

The complementary chemical structure of DNA allows, in principle, the possibility of self-duplication by a mechanism which conserves but separates each strand, with reproduction of a molecular 'progeny' of the same nucleotide sequence as the original DNA. Direct evidence for such a 'semi-conservative' complementary mechanism has been obtained for the replication of DNA in bacteriophage, bacteria and mammalian cells. The replicating unit has been proved in some cases to be a double-helical molecule of DNA, although there are indications that DNA from proliferating cells might exist as two double helices associated laterally and may thus replicate. Replication of DNA occurs with negligibly small errors in reproducing a nucleotide sequence. There is little definite information on the unwinding time of DNA and whether or not the two strands separate fully during duplication (Chapter 8).

The modification of nucleic acid structure

The effects of various chemical and physical agents on the double-helical structure of DNA have been studied to determine the molecular basis of hereditary changes and to obtain further knowledge of nucleic acid structure and stability. Such agents can rupture internucleotide phospho-ester linkages, covalent bonds within the nucleotides or the complementary hydrogen bonds, with loss of the double-helical configuration, or can introduce covalent cross-links between the chains. Since these effects occur in various combinations and the biological effects often involve several stages, it is not always easy to obtain a direct correlation between structural change and observed biological event. However, the gap is narrowing and it is now possible, for example, to detect recombination between sites on the chromosomes of certain bacteriophage which are so closely linked that the physical separation between the sites on the DNA must correspond to only a few nucleotide pairs (Chapter 9).

The control of protein synthesis by nucleic acids

Early studies of biochemical genetics showed that in higher organisms mutation of specific genes led to a block in the formation of specific biochemical compounds. Further exploration of the mechanism of gene action followed the introduction of micro-organisms for use in genetic investigations. Many cases are now known in which mutant strains of micro-organisms lack an enzyme activity present in wild-type strains. In other examples mutant strains have been found to produce enzyme which differs from the normal enzyme in some physical property or to produce a protein which lacks enzymic activity but is immunologically similar to an enzyme present in the wild-type strain. The first correlation between gene mutation and a change in the amino acid sequence of a protein resulted from studies of the biochemistry and genetics of sickle cell anaemia. Much more extensive information regarding the relation between genes and the amino acid sequence of a specific protein is available for the enzyme tryptophan synthetase of *E. coli*. This evidence, with similar information from other micro-organisms, shows beyond doubt that the function of one class of genes is to determine the sequence of amino acids in a specific protein. In addition to this there is evidence that, at least in bacteria, genes with regulating or control functions exist. Control genes determine whether or not particular structural genes will operate and also control the rate at which information is transferred to the centres of the cell where protein is formed (Chapter 10).

The mechanism of protein synthesis

Although the synthesis and structure of proteins is genetically controlled the site of protein synthesis is the ribosome, a cytoplasmic particle. Ribosomes synthesize proteins only when combined with messenger RNA, which is regarded as the primary product of structural genes. Evidence that the base sequence of mRNA corresponds to that of DNA has been obtained. mRNA acts as a template ensuring the correct sequence of amino acids in the protein being synthesized. The initial stage in protein biosynthesis is the activation of amino acids by activating enzymes which attach the amino acids to S-RNA. The amino acids necessary for protein synthesis reach the ribosomes attached to this S-RNA. Activating enzymes and S-RNA show a specificity for particular amino acids. The S-RNA molecules are believed to contain a coding segment which allows them to recognize and pair with the corresponding segment of mRNA thus ensuring that the amino acids are aligned in the correct order for the protein being synthesized. Proteins are synthesized by the step-wise addition of single amino acid residues beginning from the amino-terminal end (Chapter 11).

The genetic code

There is positive evidence that the sequence of nucleotide pairs in DNA carries the genetic code, that genetic information is passed on to RNA and that this information is translated into amino acid

169

sequence in proteins. The 'coding problem' concerns the relation between sequences of nucleotides in DNA and of amino acids in proteins. The experimental evidence shows that the code is universal, partially degenerate, is not overlapping and is read from a fixed starting point. A sequence of three nucleotides is the code for a given amino acid and many of these codes have been determined. The mechanism of transfer of genetic information in the form of nucleotide sequences from DNA to RNA has not been determined but hybrid double helices, containing one strand of DNA and one of RNA, are thought to be involved. Molecular mechanisms for variation within the DNA structure have been devised to explain genetical substitution, deletion and mutation (Chapter 12).

ADDENDUM

The suggestion that in higher organisms DNA is normally found only within the nucleus has been refuted by the demonstration that small amounts of DNA are present in a number of other organelles[1], e.g. mitochondria, chloroplasts, kinetosomes, kinetoplasts, centrioles. All of these organelles appear to be self-replicating, i.e., they do not arise *de novo* but by division of a pre-existing structure. The base ratio of DNA from the cytoplasmic organelles differs significantly from that of the corresponding nuclear DNA. There is evidence to indicate that the DNA present in mitochondria, chloroplasts and kinetoplasts is synthesized in the organelle and not in the nucleus. While there is yet no direct demonstration that the DNA from cytoplasmic organelles is involved in the synthesis of specific proteins it is becoming increasingly apparent that such DNA plays an important role in the genetic continuity of the organelle.

In autoradiographic studies of *E. coli*, labelled with tritiated thymidine, Cairns[2] has shown that the bacterial chromosome comprises a single piece of DNA which is probably duplicated at a single growing point. Direct evidence that this DNA is in the form of a circle, while it is being replicated, has now been obtained by autoradiography of cells lysed with lysozyme[3]. The total length of the chromosome seen on the autoradiographs was $1,100\,\mu$, equivalent to a molecular weight of $2 \cdot 8 \times 10^9$, which is the value expected for one continuous DNA molecule comprising the total cell content of 4×10^9 Daltons, when this is multiplied by *ln* 2 to correct for continuous duplication. This recalls the earlier hypothesis of a circular genetic map which was based on studies of recombination in *E. coli*[4] and a similar model for chromosome structure[5]. There is good physical evidence that the virus ϕ X-174 in its replicating double-helical form is circular[6] and the electron microscope also reveals ring forms with other viruses[7]. Genetic evidence for circularity has been amplified by studies on T4 phage[8] and on λ phage[9]. Examination of DNA extracted from mitochondria of mouse fibroblasts[10] indicates that it is primarily, but probably not exclusively, circular. Circular DNA may also be present in mitochondria from other cell types. The physical properties, and in particular the molecular weights, of DNAs have been surveyed[10a].

171

Closer examination has been made of the nucleotide sequences in calf thymus and herring sperm DNA; sequences of up to 7 consecutive thymidine residues have been detected[11]. Glucose is attached to the hydroxymethyl cytosine of different T-even phages to different extents and the site of attachment appears to be restricted to one glucose in sequences of two or three adjacent hydroxymethyl cytosine residues[12]. Sequence studies in nucleic acids have been reviewed[12a].

Soluble RNA can hybridize with DNA and shows saturation with excess RNA. It has been suggested that there may be as many as 40 different S-RNA molecules, coded by DNA[13]. The RNA from 18S and 28S ribosomes have been shown to have different base sequences through hybridization studies with DNA[14]. The problem of separation of pure S-RNA fractions is, of course, closely related to the question of degeneracy[15].

Trosko and Wolff[16] have observed four-stranded chromatids following treatment of isolated *Vicia faba* chromosomes with trypsin. In addition, chromosomes treated with trypsin appeared to be fragmented. Other enzymes (pepsin, DNAase and RNAase) had no gross effect on the morphology of the chromosomes. On the basis of this evidence the mitotic chromosomes of *V. faba* appears to be multistranded and depends on protein for its linear integrity. The study of honeybee and human chromosomes has lead Du Praw[17], on the other hand, to suggest that these chromosomes contain only a single strand of DNA held in a regular secondary helix by a protein sheath. While the results obtained with *Vicia* chromosomes confirm previous observations made with the light microscope they are not consistent with the results obtained by Gall[18] or Callan and MacGregor[19] following treatment of lampbrush chromosomes with enzymes.

Some of the evidence on the structure of chromosomes has been reviewed by Whitehouse[20] who favours the view that each chromatid is a single-stranded structure. A more cautious view is that, if all chromosomes are similar in organization, there is at present no hypothesis of chromosome structure which explains satisfactorily the available cytological and genetical evidence.

The structure and role of histones in deoxyribonucleohistone have been attracting more attention. The fractionation of histones is becoming increasingly exact[21] and the report of the first conference entirely devoted to histones and nucleohistones has been published[22]. The interest of such studies has been enhanced by observations that the various histone fractions inhibit DNA in its priming action on RNA polymerase to different extents[23] and that this is correlated with their ability to stabilize DNA to thermal

172

denaturation[24]. The results of optical and infra-red studies on the structure of nucleohistones have been surveyed[25] and various proposals involving cross-linking of histone between different DNA molecules have been made[26]. However, it now seems probable that the x-ray diffraction pictures of nucleohistone fibres for different moisture contents are best understood, when allowance is made for the presence of lipid, in terms of separate discrete molecules of deoxyribonucleohistone (one DNA molecule with histone longitudinally disposed along it), separated by water and not cross-linked by histone[27]. In the fibre, these molecules take the form of 'supercoils' in which the axis of the DNA helix is itself coiled with a pitch of 120 Å and a diameter of 100 Å.

Bonner and his colleagues have demonstrated[28] that in pea buds histone acts as a repressor of the synthesis of a specific protein. Vegetative pea buds do not synthesize pea seed globulin *in vivo*. Chromatin extracted from the cells of pea buds does not support the synthesis of pea seed globulin in an *in vitro* system although other proteins are formed. If, however, one fraction of the histone (I) is removed from the chromatin before it is added to the *in vitro* system then about 2 per cent of the soluble protein formed is pea seed globulin. Removal of all the histone gives a smaller proportion of globulin in the protein synthesized. Thus it appears that histone I selectively represses some of the genes in the pea genome including the genes involved in pea seed globulin synthesis. The possible role of histones in gene regulation was the subject of a recent symposium[29]. Various aspects of the structure and function of histones have been surveyed in publications by Murray[30] and Busch[31].

The study of the replication of DNA has been advanced by detailed examination of the conditions governing the synthesis of DNA by polymerases under the control of DNA primer. The primer DNA, not the enzyme, determines which bases in a mixture are assembled into the DNA chain[32]. Evidence that the primer DNA actually acts as a template and, furthermore, that one strand serves as primer and the other as template has been obtained in experiments in which (1) a DNA phosphatase-exonuclease was used to degrade native, labelled DNA from a free 3'-hydroxyl end on one of the strands; (2) the DNA was then 'repaired' with the polymerase system; and (3) the products and primer were compared by density gradient centrifugation[33].

It now appears that the product of extensive synthesis differs from DNA isolated from nature in exhibiting marked branching in electron micrographs and in renaturing readily after thermal or alkaline denaturation[33] (Richardson, C. C., Schildkraut, C. L. and Kornberg, A.) Indeed, it does not appear to undergo strand separation at all[34].

The structural requirements in DNA needed for it to act as a primer in the DNA polymerase system are certainly more subtle and complex than at first seemed likely[35]. Autoradiography of labelled DNA of *E. coli* has shown intermediate forms during replication which are those expected for semi-conservative replication of a circular double helix[36].

Convincing evidence that the sequence of amino acid residues in a protein is specified by the sequence of nucleotides in a gene has been obtained for the tryptophan synthetase system in *E. coli*[37]. Sixteen mutants with alterations in one segment of the A gene and the A protein of tryptophan synthetase were examined. From the results it is clear that the positions of the amino acid replacements in the A protein are in the same relative order as the mutationally altered sites of the corresponding mutants in the A gene. Further support for the concept of a collinear relation between gene structure and protein structure has been obtained in studies of *amber* mutants which affect the head-protein of the bacteriophage T4D[38]. *Amber* mutants do not synthesize complete head-protein molecules. Only fragments of the polypeptide chain are formed and the length of the fragment depends on the mutant concerned. The fragments can be arranged in order of increasing size and analysis shows that each fragment contains all the peptides present in smaller fragments as well as some extra peptides. It has been shown that the order of the mutant sites within the head protein gene is identical to that obtained by arranging the corresponding polypeptide fragments in order of increasing size, which indicates colinearity of gene structure and protein structure.

It has been known for some time that proteins are synthesized on ribosomes but it has become apparent that the process involves not a single ribosome but a group of ribosomes associated with the same molecule of mRNA (a polyribosome or polysome).

The role of polysomes in protein synthesis is illustrated in *Figure 1*. A ribosome becomes attached to one end of an mRNA molecule and a polypeptide molecule is initiated on the ribosome. As the ribosome moves along the messenger strand, amino acid residues are added sequentially to the polypeptide, the sequence of amino acids being determined by the base sequence of the mRNA. When the ribosome with its attached polypeptide chain reaches the end of the RNA molecule it is released and the polypeptide becomes detached from the ribosome. The mRNA molecule is regarded as having several ribosomes migrating along it at the same time each carrying a polypeptide chain at a different stage of completion. The function of polysomes in protein biosynthesis has recently been reviewed by Singer and Leder and by Schweet and Heintz[39].

ADDENDUM

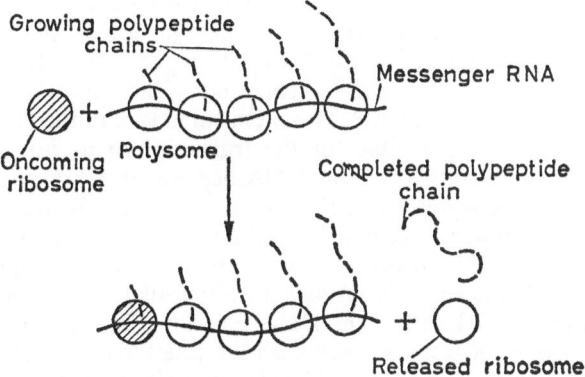

Figure 1.

A major stumbling block in the identification of an RNA fraction as mRNA is that none of the criteria used in its identification is unique to mRNA. Despite this, considerable progress has been made in elucidating a number of problems concerning the characterization, transcription and translation of mRNA.

When it became apparent that the process of protein synthesis involved not a single ribosome but a group of ribosomes (polysomes), it was generally assumed that mRNA was present in these structures. Extensive evidence is now available supporting this claim. Electron microscopic examination of polysomes shows that the ribosomes are arranged in a linear fashion joined by a thin thread whose diameter is consistent with it being RNA[40]. Polysomes are rapidly broken down to single ribosomes by treatment with a low concentration of ribonuclease which suggests that in the polysome the ribosomes are held together by a strand of RNA. Pulse labelling experiments with yeast, animal and bacterial cells also support the claim that mRNA is present in polysomes. Following brief exposure of the cells to radioactive precursors of RNA the highest specific activity is associated with the polysomes although ribosomal RNA itself has only a low specific activity. Hybridization of pulse-labelled RNA from polysomes with homologous DNA is not affected by the addition of unlabelled ribosomal RNA[41]. It has also been demonstrated in a wide variety of systems using radioactive amino acids, that nascent protein is attached to polysomes[42]: for example, the β-galactosidase of *E. coli* is attached to polysomes before it appears free[43]. This participation of ribosomes in protein synthesis has been used to support the contention that mRNA is a component of polysomes.

The genetic information present in an operon is expressed in a polarized form beginning at the operator gene end of the operon[44]. This suggests that transcription of mRNA is also polarized. Studies of the tryptophan operon of *E. coli* indicate that transcription starts at the operator end of the operon[45]. The chemical nature of the binding site of the replicase responsible for the transcription is, however, unknown. *In vitro* studies of the DNA-dependent RNA polymerase reaction, using enzyme from *E. coli*, show that the RNA formed is built up in a stepwise fashion by addition of ribonucleotides to the free 3'–OH end[46,51]. Purine nucleosides are found preferentially at the 5'–OH end. If the copying process is antiparallel, then transcription starts at the 3'–OH end of the DNA template. In the translation of mRNA base sequences into amino acid residues in protein, reading of the mRNA starts from the 5'–OH end and proceeds to the 3'–OH end. The first clear evidence on this point was obtained[47] using a synthetic polyribonucleotide ApApAp . . . C which, when added as messenger to a cell-free protein synthesizing system, resulted in the production of a polylysine with asparagine at the carboxyl end of the polypeptide:

$$H_2N-Lys-Lys-Lys . . . Lys-AspN-COOH$$

Study of the amino acid sequence of the anomalous peptides from the lysozyme of a wild type and a pseudowild double mutant strain of phage T4 also substantiate that the reading of mRNA is from the 5'–OH end to the 3'–OH end[48]. It is possible to select from the postulated codons[49,50] a unique sequence which will code for the amino acids present in the peptide from wild type lysozyme and which by deletion of one base, a shift in the reading frame and the insertion of one base, will also code for the amino acid sequence in the peptide derived from the double mutant strain (*Figure 2*). A unique base sequence coding for the two peptides can be constructed in the manner outlined only if the triplets are read from the 5' to the 3' end. In addition, these experiments confirm that the genetic code is translated by sequential reading of triplets of bases from a fixed starting point, demonstrate that proflavine can cause mutation by the addition or deletion of single bases in DNA and confirm a number of the codons proposed by Nirenberg[49,50] and Khorana[56] (*See* Table 1).

In this context the organization of the chromosome is important. If only one strand of the DNA is transcribed and the replicases of a cell have the same specificity, i.e. start at the same end of the DNA strand (present evidence favours starting at the 3'–OH end of the template[46,51]) then all operons would be expected to be oriented in the same direction. In *Salmonella* it is known that operons differ in

Wild type

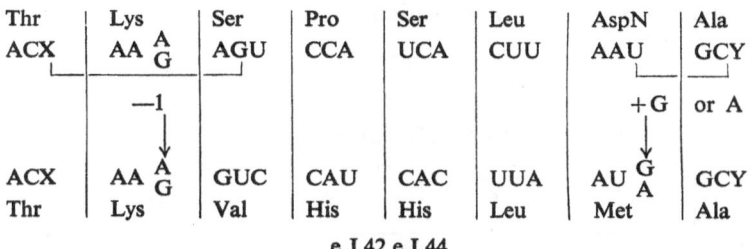

Figure 2. Amino acid sequence and related triplets for anomalous peptides from wild type and mutant phage T4 lysozyme. The 5' end of the RNA and the NH₂ end of the peptide are at the left of the figure

polarity[52] and it has been shown by transformation experiments in *B. subtilis* that linked genes on opposite strands of DNA are expressed in the cell. Thus it may be that different strands of the bacterial chromosome are transcribed in different regions. Evidence is available showing that entire operons are transcribed as a unit in the case of the histidine operon in *S. typhimurium*[53] and the tryptophan operon in *E. coli*[45]. Transcription starts at the operator end of the operon. Formation of a single mRNA molecule from an operon could provide a convenient mechanism for co-ordinate repression.

In the 1963 *Cold Spring Harbor Symposium* the results of the two main laboratories which were working on protein synthesis controlled by synthetic polyribonucleotides were surveyed[54] and more than 50 of the 64 possible nucleotide triplet codes were assigned to specific amino acids. The synthetic polyribonucleotides used were randomly ordered and thus direct determination of the sequence of the bases in the codons was not possible. More recently, Nirenberg and Leder[55] showed that trinucleotides of known base sequence promoted the binding of specific amino-acyl-RNAs to ribosomes in an *in vitro* system. The activated amino acids are not incorporated into protein under these conditions. In a series of papers Nirenberg's group and that of Khorana have each reported the results of testing the binding capacity of all 64 possible trinucleotides[50, 55, 56] derived from the bases adenine, guanine, cytosine and uracil. Table 1 shows the present best version of the genetic code based largely on the essentially similar results of the two groups. Some assignments which could not be made on the basis of binding experiments have been made using synthetic polyribonucleotides of known repeating sequencies, a technique devised by Khorana[57]. It is possible that trinucleotides which do not stimulate binding are not readable as terminal triplets in mRNA but may be readable when in an internal position.

TABLE 1

The Genetic Code

SECOND LETTER

		U	C	A	G	
F I R S T **L E T T E R**	**U**	UUU ⎫ UUC ⎭ Phe UUA ⎫ UUG ⎭ Leu	UCU ⎫ UCC UCA ⎬ Ser UCG ⎭	UAU ⎫ UAC ⎭ Tyr UAA OCHRE UAG AMBER	UGU ⎫ UGC ⎭ Cys UGA ? UGG Try	U C A G
	C	CUU ⎫ CUC CUA ⎬ Leu CUG ⎭	CCU ⎫ CCC CCA ⎬ Pro CCG ⎭	CAU ⎫ CAC ⎭ His CAA ⎫ CAG ⎭ GluN	CGU ⎫ CGC CGA ⎬ Arg CGC ⎭	U C A G
	A	AUU ⎫ AUC ⎬ Ileu AUA ⎭ AUG Met	ACU ⎫ ACC ACA ⎬ Thr ACG ⎭	AAU ⎫ AAC ⎭ AspN AAA ⎫ AAG ⎭ Lys	AGU ⎫ AGC ⎭ Ser AGA ⎫ AGG ⎭ Arg	U C A G
	G	GUU ⎫ GUC GUA ⎬ Val GUG ⎭	GCU ⎫ GCC GCA ⎬ Ala GCG ⎭	GAU ⎫ GAC ⎭ Asp GAA ⎫ GAG ⎭ Glu	GGU ⎫ GGC GGA ⎬ Gly GGG ⎭	U C A G

T H I R D   L E T T E R

Examination of the code shows that the 20 amino acids are not allotted at random to the 64 codons. The degeneracy of the code follows a pattern in which the 3′ terminal base of synonyms shows the greatest variation. U and C, A and G, or any of the four bases in the 3′ terminal position code for the same amino acid. Amino acids which are related, either structurally or metabolically, often have similar codons. This may minimize the effect of errors occurring during protein synthesis *in vivo*.

Whether or not there is a specific transfer RNA (S-RNA) for each synonym is uncertain. In some cases, e.g. serine, the number of known S-RNAs specific for the amino acid is less than the number of synonyms and thus at least some of the S-RNAs, can presumably, recognize more than one codon. It has also been shown that a single phe-RNA from *E. coli* recognizes both suggested phenylalanine codons[58]. In contrast, study of haemoglobin synthesis *in vitro* suggests that different leucyl-RNAs are required for the incorporation of leucine into different parts of the haemogblin α-chain[59].

The currently accepted concepts of the genetic code require that translation of mRNAs be started at specific points corresponding to the amino terminal ends of proteins. There is little information available at present on the initiation of translation. Experiments with synthetic polyribonucleotides indicate that primers with 5′-phosphomonoester ends are more active than those with non-esterified 5′ ends and that polyribonucleotides with 5′–purines are more active than those with 5′-pyridimidines[60]. Recent studies[61] of polypeptide chain initiation in an *E. coli* cell-free system indicate that for many (all?) peptides in this organism N-formyl-methionyl-RNA is the transfer RNA which provides the amino terminal amino acid. Since the formyl group is normally absent from finished proteins it is thought that it is removed *in vivo*. It is also suggested that in some cases the terminal methionine residue is cleaved from the polypeptide leaving the aminoterminal ends normally found in *E. coli* proteins. Clark and Marcker[62] have shown that there are at least two species of transfer RNA which can accept methionine in *E. coli* and that probably only one of these can be formylated and initiate peptide synthesis.

The biochemistry of polypeptide chain termination is as obscure as that of chain initiation. During its synthesis the growing polypeptide is attached to the S-RNA of the most recently incorporated amino acid and on completion of the polypeptide the S-RNA of the carboxy-terminal amino acid must be cleaved to release the free polypeptide. There is considerable evidence[63], from the study of certain mutants and revertants derived from them, that the information for the termination of polypeptides is coded in mRNA by the triplets UAG (Amber) and UAA (Ochre), the 'nonsense' codons. One mechanism which has been proposed is that the 'nonsense' codons correspond to an S-RNA which does not form an aminoacyl complex and that the completed polypeptide is released, in some unspecified

179

way, due to an abortive attempt to form a peptide bond with the uncharged RNA.

The universality of the code determined with the *E. coli* system has been increasingly supported by studies of other cell-free systems from mammalian and from microbial sources[64]. However, not all organisms are able to translate all code triplets[65].

The use of chemically induced mutants of the RNA from tobacco mosaic virus to produce altered virus proteins *in vivo* has been reviewed[66]. The amino acid replacements in this protein are consistent with a non-overlapping degenerate triplet code. Mutational alterations in the A proteins of the enzyme tryptophan synthetase of *E. coli* have also been surveyed[67]. In most cases change in a single nucleotide was probably responsible for each amino acid replacement.

The mechanism of the transference of information from DNA to RNA, the problem of 'transcription', has been greatly elucidated by more detailed knowledge of the reaction in which DNA primes the synthesis of RNA by means of RNA-polymerase from *E. coli*[68]. In cell-free systems, the RNA formed is complementary to both strands of the DNA primer, but recent experiments show that *in vivo* only one DNA strand is copied, that is, the mechanism is semi-conservative. This has been shown with the induction of RNA formation in *B. megatherium* by the DNA of α-phage[69], with the phage φ X-174 in its replicating form[70], and with the DNA of the SP-8 phage of *B. megatherium*[71]. It has now been shown that, in the case of the DNA of φ X-174, the transcription of one strand of the DNA into RNA *in vivo* requires that the molecule be intact and in its circular double-helical replicating form[72]. Degraded molecules, not circular and possessing two ends, each with two chain endings, were shown to replicate both strands which shows that the 'replicase' works only from one end of the DNA chains and explains the earlier apparent contrast between observations on cell-free systems and *in vivo*.

REFERENCES

[1] Granick, S. *Science, N.Y.*, 1965, **147**, 911; Corneo, G., Sanadi, D. R., Grossman, L. I. and Marmur, J., *Science, N.Y.*, 1966, **151**, 687; Luck, D. J. L. *J. Cell Biol.*, 1965, **24**, 461; Sinclair, J. H. and Stevens, B. J. *Proc. nat. Acad. Sci., Wash.*, 1966, **56**, 508; Gibor, A. and Granick, S. *Science, N.Y.*, 1964, **145**, 890; Scher, S. and Collinge, J. C. *Nature, Lond.*, 1965, **203**, 828; DuBuy, H. G., Mattern, C. F. T. and Riley, F. L. *Science, N.Y.*, 1965, **147**, 754

[2] Cairns, J. *J. mol. Biol.*, 1963, **6**, 208

[3] Cairns, J. *Cold Spr. Harb. Symp. quant. Biol.*, 1963, **28**, 43

[4] Jacob, F. and Wollman, E. L. *C.R. Acad. Sci., Paris*, 1957, **245**, 184

REFERENCES

[5] Stahl, F. W. *Proc. 11th Ann. Reunion Soc. de Chimie Phys.* Oxford; Pergamon Press, 1962

[6] Kleinschmidt, A. K., Burton, A. and Sinsheimer, R. A. *Science*, 1963, **142**, 961; Burton, A. and Sinsheimer, R. A. *Science*, 1963, **142**, 962; Chandler, B., Hayashi, M., Hayashi, M. N., and Spiegelman, S. *Science*, 1964, **143**, 47

[7] Dulbecco, R. and Vogt, M. *Proc. nat. Acad. Sci., Wash.*, 1963, **50**, 236; Weil, R. and Vinograd, J. *Proc. nat. Acad. Sci., Wash.*, 1963, **50**, 730; Ris, H. and Chandler, B. L. *Cold Spr. Harb. Symp. quant. Biol.*, 1963, **28**, 1

[8] Edgar, R. S. and Epstein, R. H. Quoted in *Ann. Rev. Microbiol.*, 1963, **17**, 104

[9] Campbell, A. *Virology*, 1963, **20**, 344

[10] Sinclair, J. H. and Stevens, B. J. *Proc. nat. Acad. Sci., Wash.*, 1966, **56**, 508

[10a] Josse, J. and Eigner, J. *Ann. Rev. Biochem.*, 1966, **35**, 789

[11] Petersen, G. B. *Biochem. J.*, 1963, **87**, 495

[12] Burton, K., Lunt, M. R., Petersen, G. B. and Siebke, J. C. *Cold Spr. Harb. Symp. quant. Biol.*, 1963, **28**, 27

[12a] Burton, K. *Essays in Biochemistry*, Vol. 1, p. 57, Ed. Campbell and Greville, Acad. Press, 1965; RajBhandary, U. L. and Stuart, A., *Ann. Rev. Biochem.*, 1966, **35**, 759

[13] Goodman, H. M. and Rich, A. *Proc. nat. Acad. Sci., Wash.*, 1962, **48**, 2101

[14] Chipcase, M. I. H. and Birnstiel, M. L. *Proc. nat. Acad. Sci., Wash.*, 1963, **50**, 1101

[15] Goldstein, J., Bennett, T. P. and Craig, L. C. *Proc. nat. Acad. Sci., Wash.*, 1964, **51**, 119

[16] Trosko, J. E. and Wolff, S. *J. cell. Biol.*, 1965, **26**, 125

[17] DuPraw, E. J. *Nature, Lond.*, 1966, **209**, 577

[18] Gall, J. G. *Nature, Lond.*, 1963, **198**, 36

[19] Callan, H. G. and MacGregor, H. C. *Nature, Lond.*, 1958, **181**, 479

[20] Whitehouse, H. L. K. *The Mechanism of Heredity*, London; Edward Arnold, 1965

[21] Rasmussen, P. S., Murray, K. and Luck, J. M. *Biochemistry*, 1962, **1**, 79; Phillips, D. M. P. *Progr. Biophys.*, 1962, **12**, 211

[22] Bonner, J. and Ts'o, P. *The Nucleohistones.* San Francisco; Holden-Day Inc., 1964

[23] Huang, R. C. and Bonner, J. *Proc. nat. Acad. Sci., Wash.*, 1962, **48**, 1216; Bonner, J., Huang, R. C. and Gilden, R. V. *Proc. nat. Acad. Sci., Wash.*, 1963, **50**, 893; Barr, G. C. and Butler, J. A. V. *Nature, Lond.*, 1963, **199**, 1170

[24] Huang, R. C., Bonner, J. and Murray, K. *J. mol. Biol.*, 1964, **8**, 54

[25] Bradbury, E. M. and Crane-Robinson, C. In *The Nucleohistones*, p. 117, (Ed. Bonner and Ts'o). San Francisco; Holden-Day Inc., 1964

[26] Bonner, J. and Ts'o, P. *The Nucleohistones*, pp. 95, 108. San Francisco; Holden-Day Inc., 1964

[27] Pardon, J. F., Wilkins, M. H. F. and Richards, B. M. *Nature, Lond.* 1967, **215**, 500

[28] Bonner, J., Huang, R. C. C. and Gilden, R. *Proc. nat. Acad. Sci., Wash.*, 1963, **50**, 893

[29] de Reuck, A. V. S. and Knight, J. *Histones: their role in the transfer of genetic information*, Ciba Fdn. Study Gp. 24, London; Churchill, 1966

[30] Murray, K. *Ann. Rev. Biochem.*, 1965, **34**, 209

[31] Busch, H. *Histones and other Nuclear Proteins*, London and New York; Academic Press, 1965

[32] Okazaki, T. and Kornberg, A. *J. Biol. Chem.*, 1964, **239**, 259

[33] Richardson, C. C., Schildkraut, C. L. and Kornberg, A. *Cold Spr. Harb. Symp. quant. Biol.*, 1963, **28**, 9; Richardson, C. C., Schildkraut, C. L., Aposhian, H. V. and Kornberg, A. *J. Biol. Chem.*, 1964, **239**, 222; Lehman, I. R. and Richardson, C. C. *J. Biol. Chem.*, 1964, **239**, 233; Richardson, C. C. and Kornberg, A. *J. Biol. Chem.*, 1964, **239**, 242; Richardson, C. C., Lehman, I. R. and Kornberg, A. *J. Biol. Chem.*, 1964, **239**, 251

[34] Bollum, F. J. *Cold Spr. Harb. Symp. quant. Biol.*, 1963, **28**, 21

[35] Bollum, F. J. *Progr. Nucleic Acid Res.*, 1963, **1**, 1

[36] Cairns, J. *Cold Spr. Harb. Symp. quant. Biol.*, 1963, **28**, 43

[37] Yanofsky, C. *Cold Spr. Harb. Symp. quant. Biol.*, 1963, **28**, 581; Yanofsky, C., Carlton, B. C., Guest, J. R., Helinski, D. R. and Henning, U. *Proc. nat. Acad. Sci., Wash.*, 1964, **51**, 266

[38] Sarabhai, A. S., Stretton, A. O. W., Brenner, S. and Bolle, A. *Nature, Lond.*, 1964, **201**, 13

[39] Singer, M. F. and Leder, P. *Ann. Rev. Biochem.*, 1966, **35**, 195; Schweet, R. and Heintz, R. *Ann. Rev. Biochem.*, 1966, **35**, 723

[40] Mathias, A. P., Williamson, R., Huxley, H. E. and Page, S. *J. mol. Biol.*, 1964, **9**, 154

[41] Schaechter, M., Previc, E. P. and Gillespie, M. E. *J. mol. Biol.*, 1965, **12**, 119

[42] Dresden, M. and Hoagland, M. B. *Science, N.Y.*, 1965, **149**, 647; Glowacki, E. R. and Millette, R. L. *J. mol. Biol.*, 1965, **11**, 116; Marcus L., Bretthauer, R., Halvorson, H. and Bock, R. *Science, N.Y.*, 1965, **147**, 615

[43] Kiho, Y. and Rich, A. *Proc. nat. Acad Sci.*, Wash., 1964, **51**, 111

[44] Alpers, D. H. and Tomkins, G. M. *Proc. nat. Acad. Sci., Wash.*, 1965, **53**, 797; Ames, B. N. and Martin, R. G. *Ann. Rev. Biochem.*, 1964, **33**, 235; Goldberger, R. F. and Berberich, M. A. *Proc. nat. Acad. Sci., Wash.*, 1965, **54**, 279

[45] Imamoto, F., Morikawa, N. and Sato, K. *J. mol. Biol.*, 1965, **13**, 169

[46] Maitra, U. and Hurwitz, J. *Proc. nat. Acad. Sci., Wash.*, 1965, **54**, 815

[47] Salas, M., Smith, M., Stanley, W., Wahba, A. and Ochoa, S. *J. Biol. Chem.*, 1965, **240**, 3988

[48] Terzaghi, E., Okada, Y., Streisinger, G., Emrich, J., Inouye, M. and Tsugita, A. *Proc. nat. Acad. Sci., Wash.*, 1966, **56**, 500; Streisinger, G., Okada, Y., Emrich, J., Newton, J., Tsugita, A., Terzaghi, E. and Inoaye, M. *Cold Spr. Harb. Symp. quant. Biol.* 1966, **31**, 77

[49] Nirenberg, M., Leder, P., Bernfield, M., Brimacombe, R., Trupin, J., Rottman, F. and O'Neal, C. *Proc. nat. Acad. Sci., Wash.*, 1965, **53**, 1161

[50] Nirenberg, M., Caskey, T., Marshall, R., Brimacombe, R., Kellogg, D., Doctor, B., Hatfield, D., Levin, J., Rottman, F., Pestka, S., Wilcox, M. and Anderson, F. *Cold Spr. Harb. Symp. quant. Biol.*, 1966, **31**, 11

[51] Bremer, H., Konrad, M. W., Gaines, K. and Stent, G. S. *J. mol. Biol.*, 1965, **13**, 540

[52] Margolin, P. *Science, N.Y.*, 1965, **147**, 1456

[53] Ames, B. N. and Martin, R. G. *Ann. Rev. Biochem.*, 1964, **33**, 235

[54] Nirenberg, M. W., Jones, O. W., Leder, P., Clark, B. F. C., Sly, W. S. and Pestka, S. *Cold Spr. Harb. Symp. quant. Biol.*, 1963, **28**, 549; Speyer, J. F., Lengyel, P., Basilio, C., Wahba, A. J., Gardner, R. S. and Ochoa, S *Cold Spr. Harb. Symp. quant. Biol.*, 1963, **28**, 559

REFERENCES

[55] Nirenberg, M. and Leder, P. *Science, N.Y.*, 1964, **145**, 1399

[56] Khorana, H. G., Buchi, H., Ghosh, H., Gupta, N., Jacob, T. M., Kossel, H., Morgan, R., Narang, S. A., Ohtsuka, E. and Wells, R. D. *Cold Spr. Harb. Symp. quant. Biol.*, 1966, **31**, 39

[57] Khorana, H. *Fedn. Proc. Fedn. Am. Socs. exp. Biol.*, 1965, **24**, 1473

[58] Bernfield, M. R. and Nirenberg, M. W. *Science, N.Y.*, 1965, **147**, 479

[59] Weisblum, B., Gonano, F., von Ehrenstein, G. and Benzer, S. *Proc. nat. Acad. Sci., Wash.*, 1965, **53**, 328

[60] Abell, C. W., Rosini, L. A. and Ramseur, M. R. *Proc. nat. Acad. Sci., Wash.*, 1965, **54**, 608

[61] Adams, J. M. and Capecchi, M. R. *Proc. nat. Acad. Sci., Wash.*, 1966, **55**, 147; Clark, B. F. C. and Marcker, K. A. *Nature, Lond.*, 1965, **207**, 1038; Webster, R. E., Engelhardt, D. L. and Zinder, N. *Proc. nat. Acad. Sci., Wash.*, 1966, **55**, 155; Leder, P. and Burszytn, H. *Proc. nat. Acad. Sci., Wash.*, 1966, **56**, 1579

[62] Clark, B. F. C. and Marcker, K. A. *J. mol. Biol.*, 1966, **17**, 394

[63] Brenner, S., Stretton, A. O. W. and Kaplan, S. *Nature, Lond.*, 1965, **206**, 994; Bretscher, M. S., Goodman, H. M., Manninger, J. H. and Smith, J. D. *J. mol. Biol.*, 1965, **14**, 634; Ganoza, M. and Nakamoto, T. *Proc. nat. Acad. Sci., Wash.*, 1966, **55**, 162; Weigert, M. G. and Garen, A. *Nature, Lond.*, 1965, **206**, 992

[64] Arnstein, H. R. V., Cox, R. A. and Hunt, J. A. *Nature, Lond.*, 1962, **194**, 1042; Weinstein, I. B. and Schechter, A. N. *Proc. nat. Acad. Sci., Wash.*, 1962, **48**, 1686; Weinstein, I. B. *Cold Spr. Harb. Symp. quant. Biol.*, 1963, **28**, 579; Maxwell, E. S. *Proc. nat. Acad. Sci., Wash.*, 1962, **48**, 1639; Marcus, L. and Halvorson, H. O. *Fed. Proc.*, 1963, **22**, 302; Gardner, R. S., Wahba, A. J., Basilio, C., Miller, R. S., Lengyel, P. and Speyer, J. F. *Proc. nat. Acad. Sci., Wash.*, 1962, **48**, 2087

[65] Benzer, S. and Champe, S. P. *Proc. nat. Acad. Sci., Wash.*, 1962, **48**, 1114; Garen, A. and Siddiqi, O. *Proc. nat. Acad. Sci., Wash.*, 1962, **48**, 1121

[66] Wittmann, H. G. and Wittmann-Liebold, B. *Cold Spr. Harb. Symp. quant. Biol.*, 1963, **28**, 589

[67] Yanofsky, C. *Cold Spr. Harb. Symp. quant. Biol.*, 1963, **28**, 581

[68] Chamberlin, M. and Berg, P. *Proc. nat. Acad. Sci., Wash.*, 1962, **48**, 81; Wood, W. B. and Berg, P. *Cold Spr. Harb. Symp. quant. Biol.*, 1963, **28**, 237; Huang, R. C. and Bonner, J. *Proc. nat. Acad. Sci., Wash.*, 1962, **48**, 1216; Robison, M. and Guild, W. R. *Fed. Proc.*, 1963, **22**, 643; Hurwitz, J., Evans, A., Babinet, C. and Skalka, A. *Cold Spr. Harb. Symp. quant. Biol.*, 1963, **28**, 59; Hurwitz, J. and August, J. T. *Progr. Nucleic Acid Res.*, 1963, **1**, 59; Huang, R. C. and Bonner, J. In *The Nucleohistones*, p. 262 (Ed. Bonner and Ts'o). San Francisco; Holden-Day Inc., 1964

[69] Tocchini-Valentini, G. P., Stodolsky, M., Aurisicchio, A., Sarnat, M., Graziosi, F., Weiss, S. B. and Geiduschek, E. P. *Proc. nat. Acad. Sci., Wash.*, 1963, **50**, 935

[70] Chamberlin, M. and Berg, P. *Cold Spr. Harb. Symp. quant. Biol.*, 1963, **28**, 67; Hayashi, M., Hayashi, M. N. and Spiegleman, S. *Proc. nat. Acad. Sci., Wash.*, 1963, **50**, 664

[71] Marmur, J. and Greenspan, C. *Science, N.Y.* 1963, **142**, 387

[72] Hayashi, M., Hayashi, M. N. and Spiegleman, S. *Proc. nat. Acad. Sci., Wash.*, 1964, **51**, 351

GLOSSARY OF SOME BIOLOGICAL TERMS

Allel: One of a number of alternative forms of the same gene, which occupy the same relative position in homologous chromosomes.

Cistron: Two mutations are located in two different cistrons or functional units if they produce the same phenotype whether present in the *cis* or *trans* configuration in a heterozygote or heterocaryon. If the *trans* configuration produces a mutant phenotype the two mutations are located in the same cistron. In the *cis* configuration both mutations are present in the same chromosome and the homologous chromosome is wild-type. In the *trans* configuration each of the homologous chromosomes carries one of the two mutations and is wild-type at the site of the other mutation. The sequence of bases in a cistron determines the sequence of amino acids in a polypeptide. If the two mutations are present in the *cis* configuration then one chromosome is wild-type at both sites and will produce normal polypeptide. If the two mutations are present in the *trans* configuration then either (*a*) both involve the same cistron. In this case both chromosomes will produce a homologous polypeptide but with a different defect, no normal polypeptide will be synthesized and a mutant phenotype will result.

Or (*b*) each mutation is in a different cistron. Under these circumstances each of the two cistrons is represented in the organism by one wild-type and one mutant form. The organism can, therefore, produce the wild-type form of each of the polypeptides and will have a wild-type phenotype.

+ +	+ b
a b	a +
cis configuration.	*trans* configuration.

a and *b* indicate mutations at different sites in a chromosome.
+ indicates that the wild-type base sequence is present.

Codon: The smallest combination of bases in a polynucleotide that codes for one amino acid[1].

Coding ratio: The ratio of the number of bases in nucleic acids to the

[1] Crick, F. H. C. *Progr. Nucleic Acid Res.*, 1963, **1**, 166

185

number of amino acids whose sequence they determine in a suitably long polypeptide[1].

Colinearity: The amino acid sequence of a protein and the equivalent sequence of bases in a nucleic acid are said to be 'collinear' when the order of amino acids along the polypeptide chain is the same as the order of the corresponding codons along the polynucleotide chain[1, 2].

Crossing-over: Exchange of chromosome segments between homologous chromosomes.

Degeneracy: A code is said to be non-degenerate if there is only one codon for each amino acid. Otherwise it is said to be degenerate[1].

Ergosome[3]: See polysome.

Gene: A unit hereditary factor located at a specific site in the chromosome and controlling the development of a specific character in the organism. Mutation of a gene results in loss or modification of the characteristic which it governs.

Heterocaryon: An organism whose multi-nucleate cells contain nuclei of different genetic constitution (used with reference to Neurospora, for example, page 119).

Heterozygous: An individual is said to be heterozygous with respect to a particular gene if it carries different alleles of this gene in homologous chromosomes.

Homologous chromosomes: The chromosomes of diploid organisms which constitute a pair at meiosis.

Homozygous: An individual is said to be homozygous with respect to a particular gene if it carries the same allele of this gene in homologous chromosomes.

Operator gene and operon: See accounts on pp. 124–126 and *Figure 10.1.*

Overlapping: A code is said to be overlapping if a given base forms part of several codons[1].

Polysome: An aggregate of ribosomes which is active in the synthesis of proteins[4, 5] (= ergosome[3]).

Recessive allele: Alleles are said to be recessive when they are detectable only in homozygous individuals.

Universal code: A code is universal if, throughout nature, any particular codon has the same meaning[1].

[1] Crick, F. H. C. *Progr. Nucleic Acid Res.*, 1963, **1**, 166
[2] Yanofsky, C. *Cold Spring Harb. Symp. quant. Biol.*, 1963, **28**, 581; Yanofsky, C., Carlton, B. C., Guest, J. R., Helinski, D. R. and Henning, U. *Proc. Nat. Acad. Sci. U.S.*, 1964, **51**, 266
[3] Wettstein, F. O., Staehelin, T. and Noll, H. *Nature, Lond.*, 1963, **197**, 430
[4] Pirie, N. W. *J. Inst. Biol.*, 1964, **11**, 92
[5] Warner, J. R., Rich, A. and Hall, C. E. *Science*. 1962, **138**, 1399

INDEX

Acridine mutants (of bacteriophage), 150, 158
Acridines, bacteriostatic effect, 158
Adaptor hypothesis, 132, 134
Alkaptonuria, 117
Allele, 117, 185, 186
Amino acid,
  activating enzymes, 129, 130
  substitutions in proteins, 121, 122, 140, 147–150
  transfer RNA (*see* Soluble RNA)
Amylase synthesis, 136
Anthocyanins, 117
Autoradiography, 78–81, 82, 171,
  chromosome constituents, and, 75
  T2 DNA, of, 93, 94

Bacteria,
  recombination in, 18, 19
  sexual reproduction in, 18
Bacteriophage, 15–17, 123, 150–154
  DNA of, 15–17, 27, 33, 35, 37, 93, 94, 96
  replication, 93, 94
  T2, 27, 35, 93, 94
  T4, 150–153, 171, 174, 176
  φX-174, 28, 33, 35, 171
Base pairing specificity, 31, 97

Chemical mutagens (*see also* Nucleic acids), 79
Chromosome(s), 7, 72–85
  autoradiographic studies, 79–81, 171
  bacterial, 18
  chemical constituents, 72–75
  cytochemical reactions, 72
  direct chemical analysis, 73
  duplication, 79–81
  electron microscopic studies, 76
  enzymatic studies, 74
  genes, and, 4
  induced structural changes, 78
  isotopic studies, 74
  lampbrush, 76, 79, 82, 83
  lateral organization, 75, 172

Chromosome(s) (*cont.*)
  linear organization, 81
  map, 4
  models, 81–85
  multiplication, 3
  salivary gland (Drosophila), 7, 8
  spirals, 110
  structural organization, 75
  theory of inheritance, 4
Cistron, 115, 150, 185
Clover-leaf RNA, 48, 49
Clupeine, 64
Coding problem, 140, 176
Coding ratio, 142, 143, 146, 153, 154, 185
Codon, 176, 178, 185
Colchicine, 80
Colinearity, 174, 186
Crossing-over, 4, 185

Degeneracy, 143, 150, 154, 185
Density gradient centrifugation, 94
2-Deoxy-D-ribose, 24
Deoxyribonuclease, 24, 75, 79, 82, 102, 109
Deoxyribonucleohistone, 66–68, 77, 172
  molecular weight, 67
Deoxyribonucleoprotamine, 64–67
Deoxyribonucleoproteins, 7, 63–68
Deoxyribonucleic acid, DNA (*see also* Nucleic acids),
  bacterial, 18, 38
  bacteriophage (*see* Bacteriophage)
  base addition, 157, 158
  base deletion, 158
  base substitution, 157
  chemical structure, 23–26
  chromosome constituent, 72–75
  circular, 36, 171
  composition, 9, 26–28, 143
  configuration, 28–34
  content of nucleus, 7, 8, 78
  denaturation, 31, 107–111
  dissociation curves, 26, 27
  ducks, from, 19